Shaping Technology/Building Society

Inside Technology
edited by Wiebe E. Bijker, W. Bernard Carlson, and Trevor Pinch

Wiebe E. Bijker and John Law, editors, *Shaping Technology/Building Society: Studies in Sociotechnical Change*

Stuart S. Blume, *Insight and Industry: On the Dynamics of Technological Change in Medicine*

H. M. Collins, *Artificial Experts: Social Knowledge and Intelligent Machines*

Pamela E. Mack, *Viewing the Earth: The Social Construction of the Landsat Satellite System*

Donald MacKenzie, *Inventing Accuracy: A Historical Sociology of Nuclear Missile Guidance*

Shaping Technology/Building Society
Studies in Sociotechnical Change

edited by Wiebe E. Bijker and John Law

The MIT Press
Cambridge, Massachusetts
London, England

This book was set in Baskerville by Asco Trade Typesetting Ltd., Hong Kong, and printed and bound in the United States of America.

Library of Congress Cataloging-in-Publication Data

Shaping technology/building society : studies in sociotechnical change / edited by Wiebe E. Bijker and John Law.
    p. cm.—(Inside technology)
  Includes bibliographical references and index.
  ISBN 0-262-02338-5 (HB), 0-262-52194-6 (PB)
  1. Technology—Social aspects. I. Bijker, Wiebe E. II. Law, John, 1946–
III. Series.
T14.5.S49   1992
303.48′3—dc20   91-48122
                                                                     CIP

10 9 8 7 6

Contents

9
A Summary of a Convenient Vocabulary for the Semiotics of Human and Nonhuman Assemblies
Madeleine Akrich and Bruno Latour

10
Technology, Testing, Text: Clinical Budgeting in the U.K. National Health Service
Trevor Pinch, Malcolm Ashmore, and Michael Mulkay

11
Postscript: Technology, Stability, and Social Theory
John Law and Wiebe E. Bijker

Preface

Earlier versions of many of the papers in this volume were first presented at a workshop in The Netherlands in 1987. The object of that workshop was to explore the intriguing place where history meets sociology in the social analysis of technology. Specifically, the concern was to consider the extent to which case studies can be used to build or test theory about the way in which the "seamless web" of the social and the technical is structured and shaped. Most of the papers in this volume (like others presented at the workshop) wrestle with the interaction between theory and data by exemplifying ways in which the two might be brought together.

We are grateful for the help we received from colleagues who acted as referees to the individual papers. Robert Frost made extremely valuable comments on each paper and on the volume as a whole. We are most grateful to him.

General Introduction

Wiebe E. Bijker and John Law

What Catastrophe Tells Us about Technology and Society

On Monday, October 16, 1989, two of the contributors to this volume were driving along Interstate 880 through Oakland, California. The highway was a road, just another road, scarcely worthy of attention. Insofar as they thought about it at all, they wrestled with traffic conditions and route-finding. For instance, there was a tense moment when they changed lanes—they really didn't want to be forced onto the Bay Bridge and end up in San Francisco. But once the problem of route-finding was solved, they scarcely noticed as they drove onto the Nimitz Highway section of the road. Carried along at rooftop level like the other motorists, they continued their conversation; the history of the freeway and its mode of construction were not issues that concerned them at all.

The next day, at 5:04 P.M., deep beneath the Santa Cruz mountains, the strains caused by the grinding of two tectonic plates reached breaking point. The earth shook, and sixty miles to the north the shock waves reached the Bay area. Buildings swayed, gas mains ruptured, landfill rocked, and the Cypress Structure section of the Nimitz Highway collapsed. Those who were lucky were driving on the top deck, or they found an exit waiting for them and drove down a ramp. Or like our two contributors and millions of others, they had passed along the highway before the shock struck. Those who were unlucky found themselves trapped between the two decks. Some were rescued. Tragically many, as we know, were killed.

Most of the time, most of us take our technologies for granted. These work more or less adequately, so we don't inquire about why or how it is they work. We don't inquire about the design decisions that shape our artifacts. We don't think very much about the ways in which professional, political, or economic factors may have given form to those designs—or the way in which they were implemented in practice. And even when our technologies do go wrong, typically our first instinct is to call the repair person. There are routine methods for putting them right: it doesn't occur to us to inquire deep into their provenance.

In one sense, our lack of curiosity makes perfect sense. If we stopped to think why our artifacts—our saucepans, our cars, our refrigerators, our bridges—work or take the form that they do, we would never get around to boiling the water to make coffee each morning. The conduct of daily life surely demands a *tactical* lack of curiosity! But that lack of curiosity carries costs and overhead expenses as well as benefits. Our artifacts might have been different. They might have worked better. They might not have failed. They might have been more user—or environmentally—friendly.

Most of the time the relationship between such costs and overheads on the one hand, and the benefits that accrue from technologies on the other, is well hidden. Only on occasion—when, for instance, we think about catalytic converters, the merits or drawbacks of nuclear power, or access to buildings for the physically handicapped—do we start to think about the trade-offs. And the costs of technology tend to become obvious only at moments of catastrophic failure—when we suddenly realize that, somewhere along the line, there was something lethally wrong with a technology that we were used to taking for granted. These are the moments when we learn that space shuttles, car ferries, nuclear reactors, passenger escalators, oil platforms, or other engineered constructions like the Cypress Structure of the Nimitz Highway were, as it turned out, fatally flawed.

This is a book about technologies and the ways in which they are shaped. In one sense, then, it undertakes the kind of job done by the San Francisco earthquake on the Nimitz Highway, albeit in a less destructive manner. Thus in the aftermath of that tragedy, it became clear that the concrete pillars of the Cypress Structure were reinforced with vertical rods—there were no additional spiral rods. It became clear that engineers knew that this was the case and worried about it because they thought the structure might collapse if there were an earthquake. It became clear that proposals to reinforce the structure had been postponed because of lack of funds. And it became clear that the lack of funds was a result of a complex set of political decisions about electoral priorities and attitudes to taxation. It became clear, in short, that the Nimitz Highway was (as it turned out) fatally flawed because of the way in which a set of engineering decisions embodied or mirrored a set of professional, economic, and political realities.

It is sometimes said that we get the politicians we deserve. But if this is true, then we also get the technologies we deserve. Our technologies mirror our societies. They reproduce and embody the complex interplay of professional, technical, economic, and political factors. In saying this, we are not trying to lodge a complaint. We are not proposing some kind of technological witch hunt. We are not trying to say, "If only technologies were purely technological, then all would be well." Rather, we are saying that *all* technologies are shaped by and mirror the complex trade-offs that make up our societies; technologies that work well are no different in this respect from those that fail. The idea of a "pure" technology is nonsense. Technologies always embody compromise. Politics, economics, theories of the strength of materials, notions about what is beautiful or worthwhile, professional preferences, prejudices and skills, design tools, available raw materials, theories about the behavior of the natural environment—all of these are thrown into the melting pot whenever an artifact is designed or built. Sometimes, like the Cypress Structure, the product fails in a tragic and spectacular manner. More often it works. But—and this is the basic message of this book—its working or its failing is always shaped by a wide range of disparate factors. Technologies, we are saying, are shaped. They are shaped by a range of heterogeneous factors. And, it also follows, they might have been otherwise.

They might have been otherwise: this is the key to our interest and concern with technologies. Technologies do not, we suggest, evolve under the impetus of some necessary inner technological or scientific logic. They are not possessed of an inherent momentum. If they evolve or change, it is because they have been pressed into that shape. But the question then becomes: why did they *actually* take the form that they did? This is a question that can be broken down into a range of further questions. Why did the designers think in this way rather than that? What assumptions did the engineers, or the business people, or the politicians, make about the kinds of roles that people—or indeed machines—might play in the brave new worlds they sought to design and assemble? What constraints did they think about—or indeed run into—as they built and deployed their technologies? What were the uses—or abuses—to which the technologies were put by their users once they were deployed? How, in other words, did users themselves reshape their technologies? And how did the users and their technologies shape and influence future social, economic, and technical decisions? These are the kinds of issues that concern us—issues that go to the heart of what is sometimes called "the social shaping of technology," issues that demand that we think simultaneously about the social and the technological.

Technology, History, and Social Science

This, then, is the reason why the topic covered in this book is so important. The processes that shape our technologies go right to the heart of the way in which we live and organize our societies. Understanding these processes might help us to create different or better technologies. Understanding them would allow us to see that our technologies do not necessarily have to be the way they actually are.

Immediately, however, we start to encounter difficulties. Thus our language tends to suggest that technologies *do* indeed have an inner logic or momentum of their own. We talk, for instance, of the "technological" as if it were set apart—something that may be subverted by the political or the economic as, for instance, critics of the Nimitz Highway failure tended to suggest. But we need to overcome this linguistic inertia. We need to remind ourselves that when we talk of the technological, we are not talking of the "purely" technological— that no such beast exists. Rather we are saying that the technological is social. Already, then, we find that we need to blur the boundaries of categories that are normally kept apart. There is no real way of distinguishing between a world of engineering on the one hand and a world of the social on the other.

But this is only the first difficulty, for the word "social" presents us with analogous problems. What do we mean when we write of the "social?" Do we mean "social" as in "sociological"? The answer is that we do, but only in part. For the social is not exclusively sociological. In the context of technology and its social shaping, it is also political, economic, psychological—and indeed historical. Here, then, our terminological problems shade off into issues to do with the disciplinary organization of knowledge. There is a large body of work on the economics of research and development. There is a tradition of work in political science on the relationship between technology and bureaucratic politics. There is a related literature on military procurement. There are sociological and historical studies of the links between capital formation, new technologies, and the labor process. There is feminist work on the "gendering" of technologies. There is a literature in cultural studies on the relationship between the mass media and the formation of popular culture. There is a large body of work on organizational innovation and the introduction of new technologies. There are policy-linked studies of the process of technology transfer, and sociological and historical studies of the relationship between technological innovation and the formation of world systems. There are philosophical studies of the implication of technologies, and technological modes of rationality, for the character of communication and human interaction. And, of course, there are admirable business histories that explore the relationship between research and development and the growth of large enterprises.

Technology and its shaping has to do with the historical, the economic, the political, and the psychological, as well as with the sociological. But how can we find ways of overcoming the divisions and blinkers that academic disciplines use to set themselves apart? How can we find multidisciplinary ways of talking about heterogeneity: of talking, at the same time, of social *and* technical relations, evenhandedly, without putting one or the other in a black box whose contents we agree not to explore?

It seems to us that the way has often been led by the social history of technology. For instance, many of the papers published in *Technology and Culture* in the recent past represent a substantial attempt to wrestle with one or another aspect of the heterogeneity of technology and the scientific, organizational, social, economic, and political processes that give it shape:[1] they talk, at the same time, of *both* the rivets *and* the social relations. This is a lesson, an example, that might well be learned by many of those who contribute to the various literatures mentioned above. For the historians, technology is not the "exogenous variable" of neoclassical economics. It is not the scientifically driven fact of life found in many of the otherwise revealing studies in bureaucratic politics. Neither is it driven by the unproblematic search for greater speed, power, range, or endurance that is assumed in many studies of the procurement process.

To be sure, questions have been raised about these assumptions: evolutionary economic theories of technological change do not treat the latter as an exogenous variable,[2] and there have been studies in bureaucratic politics and procurement—admittedly influenced by sociological concerns—in which careful attention is paid to technological content.[3] However, the social history of technology, at its best, locks away neither social relations nor the content of technology. It represents a substantial, sustained, and empirically grounded attempt to come to terms with technological heterogeneity. But if we may be forgiven a general claim about a discipline not our own, it seems to us that its strength lies in its relatively unsystematic character. Few if any historians would admit to the systematizing pretensions of (for instance) economists and sociologists. This means that they are less prone to systematic and blinkering idées fixes about the relations between technologies and their environments. In particular, they are less prone to treat technology (or one or another aspect of social relations) as unexamined variables. Instead, they follow the scientists and engineers [as Latour (1987) has noted, albeit in a different context] wherever the latter go: into attempts to discipline the labor force, the character of business accounting, the nature of laboratory work, the shaping of workshop skills, professional organization, methods for technological testing, the "American" system of manufacture, the manipulation of political elites, and many more. This, then, is one of the major strengths of the social history of technology—its propensity to take on board whatever appears, in light of the evidence, to be important.

What, then, is the difficulty? One answer is that there *is* no difficulty, that good history is good history, that it is driven by a rigorous concern with necessarily heterogeneous evidence, and that this is as it should be. This is a powerful and entirely tenable position, one that demands respect. Those who wish to press the integrity of history can indeed point with some justice to the blinkering effects of the concern with systematic theory that is found in such disciplines as sociology and economics.

This said, we do not need to press far into the philosophy of history to note that *no* kind of writing, social history included, is without its own set of epistemological blinkers. As is obvious, we are *all* better at seeing some things than others. Indeed, we all have our blind spots. Social historians of technology are no exception to this rule.[4] Thus they are very good at describing the messy and heterogeneous processes that drive technological change, but they are less concerned with—and perhaps not so good at—trying to detect the regularities that may underpin those processes. The result is that sometimes—we emphasize that this is only sometimes the case—those who carry prejudices of a more sociological character find it difficult to know what to make of the attractive case studies that appear in the pages of *Technology and Culture*. The issue, then, is: is it possible to draw conclusions from these studies? Is there any possibility of generalization?

Here we are on delicate and difficult ground. It is delicate because there are, of course, historians of technology who are deeply concerned with generalization. They seek to explore and develop models, for instance of the relationship between technology and the labor process, or the development and stabilization of large technological systems.[5] It is, accordingly, far from the case that all historians of technology avoid generalization or commitment to models of the social shaping of technology. Neither, as a plethora of recent case studies suggests, is it true that sociologists, economists, or political scientists are necessarily insensitive to the differences between, or indeed the idiosyncracies of, particular empirical circumstances.

But the ground is difficult for another, more fundamental reason. This is because we are involved, here, in what amounts to a tradeoff—a trade-off between following the messy story wherever it leads us on the one hand, and trying to extract, develop, or impose more general models of the course, is that if any description is a simplification—something that we *all* have to come to terms with when we start to write—then a relatively well-structured model represents a further echelon of simplification. Thus a model or a theory, whatever its form, is a kind of statement of priorities: in effect it rests on a bet that for certain purposes some phenomena are more important than others. It simplifies down to what it takes to be the essentials. And whether or not it is a satisfactory simplification, or indeed an oversimplification, is a matter of judgment and, in the last instance, a matter of personal or disciplinary taste.

The papers brought together in this book, papers written by historians, sociologists, philosophers, and political scientists, share certain commitments. One is to the heterogeneity of technology—to the study of the *content* as well as the context of technical change. A second is a commitment to wrestle with the trade-off between the exploration of messy case studies and the attempt to built somewhat more general models or ways of thinking about the social shaping of technology. Thus, although the particular solutions that they adopt differ, the authors all seek to develop empirically sensitive models for describing the ways in which technology is shaped by social, political, or economic factors.

Common Ground: Heterogeneity and Contingency

Let us back up a step. We have said that technologies are not purely technological. Instead, we have said that they are heterogeneous, that artifacts embody trade-offs and compromises. In particular, they embody social, political, psychological, economic, and professional commitments, skills, prejudices, possibilities, and constraints. We've said that if this is the case, then the technologies with which we are actually endowed could in another world have been different. And this means that the technologies that are currently in the process of being developed might, at least in principle, take a variety of different forms, shapes, and sizes.

This means, of course, that technologies do not provide their own explanation. If there is no internal technical logic that drives innovation, then technologically determinist explanations will not do. This means that we should be similarly cautious of explanations that talk of technological trajectories or paradigms. Even if we can identify a paradigm, this does not mean that we have thereby identified what it is that drives the way in which it is articulated. And even if we can observe a trajectory, we still have to ask why it moves in one direction rather than another.[6]

We are arguing, then, that these Newtonian metaphors will not do. Technologies do not have a momentum of their own at the outset that allows them, as Latour (1987) has put it, to pass through a neutral social medium. Rather, they are subject to contingency as they are passed from figurative hand to hand, and so are shaped and reshaped. Sometimes they disappear altogether: no one felt moved, or was obliged, to pass them on. At other times they take novel forms, or are subverted by users to be employed in ways quite different from those for which they were originally intended.

But if all this is the case, we have a new problem. How does it happen that technologies ever actually firm up? Why is it that they take the form they do, rather than some other shape? How do things get settled? And this is the basic problem tackled by the contributors to this volume—the reason why they are not simply concerned with the detail of their case studies but want, in addition, to explore simplifying generalizations. And although they adopt a variety of approaches, the authors share at least five assumptions.

1. They all take it as given that technological change is indeed *contingent* in the way described above: reductionist explanations, and in particular those that assume that technological change may be explained in terms of an unfolding internal logic, are avoided. In addition, many of the contributors also avoid other forms of reductionism: they take it that what we normally call "the social" or "the economic" is, like technology, both heterogeneous and emergent; and that what we normally think of as social relations are also constituted and shaped by technical and economic means. The assumption, then, is that technology, the social world, and the course of history should all be treated as rather messy contingencies. There

is no grand plan to history—no economic, technical, psychological, or social "last instance" that drives historical change.

2. The contributors also assume that technologies are born out of *conflict, difference, or resistance.* Thus most if not all of the case studies describe technological controversies, disagreements, or difficulties. The pattern is that the protagonists—entrepreneurs, industrial or commercial organizations, government bureaucracies, customers or consumers, designers, inventors, or professional practitioners—seek to establish or maintain a particular technology or set of technological arrangements, and with this a set of social, scientific, economic, and organizational relations.

However, as one would expect, such arrangements have implications—in some cases damaging implications—for other actors or arrangements. At the very least, the latter would have acted in a different manner had they been left to their own devices. The pattern, then, is that resistance is put up. The people or agencies who "feel" these damaging implications would like to see the social relations, and the technologies in which they are implicated, take some other form. This, then, accounts at least in part for the claim we made above that things might have been otherwise: they *would* have been otherwise had other plans prevailed.

3. Such differences may or may not break out into *overt* conflict or disagreement. Thus customers do not normally riot if their needs are not being met, though they do tend to take their purchasing power elsewhere. On the other hand, rival manufacturers or government departments may put up stiff resistance, and bureaucratic, political, or economic conflict may break out. At any rate—and this is the third feature that their case studies have in common—most of the contributors are concerned to map the *strategies* deployed by those involved in dispute, disagreement, or resistance.

We shall have more to say about these in the conclusion. Note, however, that such strategies are empirically varied: thus the papers range through a series of (often combined) legal, organizational, political, economic, scientific, and technical strategies. But, although they differ in specifics, the assumption is that these are designed in all cases to box in the opposition—to stop it acting otherwise, going elsewhere, or successfully stabilizing its own alternative version of technological and social relations. Accordingly, the way in which such strategies are deployed, together with their success, marks the focal point for many of the contributors.

4. Technologies, then, form part of, or are implicated in, the strategies of protagonists. But what of our original question—how do

they they firm up? Here again, a common commitment draws the contributors together: they assume that a *technology is stabilized if and only if the heterogeneous relations in which it is implicated, and of which it forms a part, are themselves stabilized.* In general, then, if technologies are stabilized, this is because the network of relations in which they are involved—together with the various strategies that drive and give shape to the network—reach some kind of accommodation.

Little can be said about this process in the abstract. It may take the form of compromise—some kind of negotiated settlement. It may look like politics, bureaucratic or otherwise. It may look like the exercise of naked power, for the contributors certainly do not take the view that accommodation, the satisfaction of all of those involved, is a necessary outcome of struggle. The differences may be expressed in or through a variety of forms, shapes, or media: words, technologies, physical actions, organizational arrangements.

5. Finally, the contributors assume that both strategies themselves and the *consequences* of those strategies should be treated as emergent phenomena. This is an important point—and it is one that takes us right to the heart of social theory. Thus it is self-evident that when two or more strategies mesh together, the end product is an emergent phenomenon: a game of chess cannot, after all, be reduced to the strategies of either one of the players. So it is with technologically relevant controversy. Whatever the system builders may wish were the case, what actually happens depends on the strategies of a whole range of other actors.

But if this is the case for the outcome of conflicts or resistances, it is also true for the strategies mounted by actors at the moment they enter the game. Again, the analogy with chess is helpful, for here, as is obvious, the strategies developed by the players are shaped by the rules of the game—the pieces, their relationships, and the possibilities they embody. Indeed, the very notion of a player— "white" or "black"—is constituted in those rules. But if the metaphor is helpful because it underlines the way in which strategies and players are themselves built up in a field of relations, then it is also somewhat misleading, because it rests of the assumption that the rules of the game are fixed before the game starts—something that is doubtful, and usually wrong, in sociotechnical analysis.

This, then, is where we confront one of the core problems of social and political theory head-on: how it is that actors (people, organizations) are both shaped by, but yet help to shape, the context in or with which they are recursively implicated. So it is that a fairly matter-of-fact and practical question—what we can say about the firming up of technologies—leads us not only into the concerns of historians but also to the heart of sociological and political theory to the problem of the social order. And, unsurprisingly, we find that the different contributors to this volume come up with a range of different responses—though, as we have already indicated, they have in common their concern to avoid reductionism, the assumption that the technical, or the social, or the economic, or the political lie at the root and so drive, sociotechnical change.

Conclusion

In this introduction we have argued that a concern with the social shaping of technology is important for a number of separate reasons. As is obvious, technology is ubiquitous. It shapes our conduct at work and at home. If affects our health, the ways in which we consume, how we interact, and the methods by which we exercise control over one another. The study of technology, then, has immediate political and social relevance. And to be sure, because technology is treated as one of the major motors of economic growth, it has similar economic and policy relevance.

Technology does not spring, *ab initio*, from some disinterested fount of innovation. Rather, it is born of the social, the economic, and the technical relations that are already in place. A product of the existing structure of opportunities and constraints, it extends, shapes, reworks, or reproduces that structure in ways that are more or less unpredictable. And, in so doing, it distributes, or redistributes, opportunities and constraints equally or unequally, fairly or unfairly.

But, although technology is important, for reasons we have indicated, its study is fragmented: there are internalist historical studies; there are economists who are concerned with technology as an exogenous variable; more productively, there are economists who wrestle with evolutionary models of technical change; there are sociologists who are concerned with the "social shaping" of technology; and there are social historians who follow the heterogeneous fate of system builders.

The last five years has seen the growth of an exciting new body of work by historians, sociologists, and anthropologists, which starts from the position that social and technical change come together, as a package, and that if we want to understand *either*, then we really have to try to understand *both*. But the chapters that make up this volume attempt to map the fertile common ground between social science and sensitive social history. Thus they go beyond a commitment to the analysis of heterogeneity, and explore vocabularies of analysis for making sense of sociotechnical stabilization. They are concerned, then, with the regularities that underpin the contingent processes that lead to relatively stable technologies, and the social stabilities in which these are implicated.

The differences among the contributors reflect, in part, simple differences in subject matters. Thus some chapters look like studies of bureaucratic politics, some like studies of technology transfer, and some like business histories—although in all cases they differ from these because of their concern with technological content. In part, the differences reflect the range of the authors' backgrounds in history, sociology, and anthropology.

In part, however, the differences also reflect (equally healthy) differences in theoretical approach. At the place where social history meets sociology there are at least three different, somewhat overlapping, and productive traditions. First, there is a version of systems theory, which was developed in the history of technology by such writers as Hughes (1983). This was originally intended to describe and account for the growth of large technical systems. Hughes's argument—which makes good sense of the growth of such technical networks as the electricity supply industry—is that the successful entrepreneurs were those who thought in system terms, not only about the technical character of their innovations but also about their social, political, and economic context. In effect, he says that entrepreneurs like Edison designed not only devices but societies within which these devices might be successfully located. This approach, which has attracted substantial attention not only from historians but also from sociologists, has influenced a number of the contributors to this volume—and perhaps in particular the work of Carlson on the cultural construction of motion pictures, though it is ironic that Carlson's study is of a case where Edison's system building broke down.

Second, there is actor-network theory. This was first developed by Callon (1980) and represents an attempt to find a neutral vocabulary to describe the actions of those who have since been called "heterogeneous engineers" (Law 1987). The idea is that such heterogeneous engineers build messy networks that combine technical, social, and economic elements. To a first approximation actor-network theory has much in common with Hughes's version of systems theory. However, unlike Hughes, Callon and his collaborators stress that the elements (including the entrepreneurs) bound together in networks are, at the same time, constituted and shaped in those networks. This means that they avoid making assumptions about a backdrop of social, economic, or technical factors: the backdrop is something that is itself built in the course of building a network—a point made by Law and Callon in their contribution to this volume. It also means that they avoid making the commonsense assumption that people, entrepreneurs, or machines are naturally occurring categories. How boundaries are drawn between (for instance) machines and people thus becomes a topic for study in its own right. This concern is clearest in the chapters by Latour and Akrich, who seek to build a rigorous semiotic vocabulary for talking, symmetrically, about people and machines.[7]

Finally, there is the social constructivist approach to technology. This is an attempt to apply recent work in the sociology of scientific knowledge to the case of technology.[8] Thus in the sociology of science it has been argued that knowledge is a social construction rather than a (more or less flawed) mirror held up to nature. It is not given to scientists by (or uniquely by) nature whose phenomena are, it is argued, always susceptible to more than one interpretation. Rather, it is better seen as a tool. Accordingly, scientific knowledge—and now, it is argued, technologies and technological practices—are built in a process of social construction and negotiation, a process often seen as driven by the social interests of participants. This approach informs, in particular, the chapters by de la Bruhèze, Misa, and Bijker: in each case these authors are concerned with "closure"—that is, the process by which conflicting groups reach (or impose) a specific outcome and so conclude the dispute.

However, although the case studies differ both empirically and theoretically—indeed diversity in terms of the latter is greater than this brief review suggests—what emerges overwhelmingly is that the case studies complement one another. They represent a resource that can be drawn on and used by anyone concerned with the shaping of the social and the technical, whatever their specific point of view. Thus the authors have not dug a set of theoretical slit trenches, but sought to develop and make use of a range of different empirically and theoretically relevant concepts. Accordingly, each chapter contributes in its own specific way to the central aim of this volume—the development of an empirically sensitive theoretical understanding of the processes through which sociotechnologies are shaped and stabilized.

Notes

1. John Staudenmaier (1985) has described the origins of SHOT and *Technology and Culture*, and there is no call to cover that ground in detail again—although it is worthwhile observing that SHOT was established because historians concerned with technology found that there was little interest in the latter, either in general history or in the history of science.

2. See Nelson and Winter 1982 and Rosenberg 1982.

3. See, for instance, MacKenzie 1990a and Mack 1990.

4. As an example, consider the question of the objects or *units* of analysis. Internalist historians of technology tend to privilege technological or scientific logics, and explore the way in which technologies are driven by or unpack these logics as they develop. Social historians tend to privilege human beings and proceed to locate these in a technical, cultural, political, or economic environment that is taken to influence technological change. Some sociologists, by contrast, treat human beings and their environment as the expression of paradigms or discourses or social organization. For further discussion of this and the general relationship between social science and history, see Buchanan 1991, Law 1991a, and Scranton 1991.

5. The most obvious references are, respectively, Noble 1977 and Hughes 1983.

6. MacKenzie (1990a) has plausibly suggested that a *belief* on the part of many that trajectories (for instance in the direction of greater missile accuracy) exist is important for technological change.

7. The notion of methodological symmetry was spelled out by Bloor (1976) in the sociology of scientific knowledge. He argued that knowledge that is taken to be true, and knowledge that is taken to be false, are both susceptible to sociological explanation, and that explanation should be in the same terms. The notion of generalized symmetry has been developed by Callon (1986a), who argues that society and nature should be described in the same terms.

8. The argument is spelled out in detail in Pinch and Bijker 1987.

I

Do Technologies Have Trajectories?

The idea that technologies have natural trajectories is deeply built into the way we talk. Almost as deep is the notion that any *individual* technology moves through a natural life cycle: from pure through applied research, it moves to development, and then to production, marketing, and maturity. As we have indicated in the introduction, many recent studies in the social history and sociology of technology suggest that these models of innovation are quite inadequate. This message is pressed home in this volume, but particularly in the three papers in this first section. These are all concerned in one way or another with the character of technological trajectories. And they are all concerned to show that there is nothing inevitable about the way in which these evolve. Rather, they are the product of heterogeneous contingency. In addition, the three papers suggest possible vocabularies for sociotechnical analysis—for making sense of the heterogeneity and contingency of technical change.

Law and Callon take the case of the TSR.2—a British military aircraft somewhat like the F111. After various vicissitudes the TSR.2 flew—in fact quite successfully—and was then cancelled. There are various ways of reading this story. It could, for instance, be treated as another example of profligate military waste, or as an example of the way in which politics can undermine decent technology. In fact, Law and Callon choose to examine the development of the project in an evenhanded manner. Yes, they say, it *is* possible to discern a trajectory for this project. But they go on to argue that there was nothing natural or inevitable about that trajectory. It was not a consequence of a naturally unfolding process of technological development; at all points it should be seen rather as a product of contingency. The result is that it twists and turns as social and technical circumstances change. Law and Callon use a network vocabulary to document the way in which the trajectory of the TSR.2 project was affected by the heterogeneous strategies of those involved. In particular, they describe the way in which the protagonists sought to give the project a degree of autonomy from its environment—a degree of insulation from some, though only some, of its contingencies.

The importance of this process of building a boundary between inside and outside—a boundary that eventually ruptured in the case of the TSR.2, with the collapse of the project—is also emphasized by Bowker. Here again, the concern is with a technological trajectory—that of the development of geophysical methods by Schlumberger. But if the development of these methods was not inevitable, then how was this achieved? Bowker argues that the company successfully mobilized a series of resources to build a version of natural and social reality within which its methods secured success. As a part of this strategy, the company Whiggishly *claimed* that its geophysical techniques were, indeed, the product of an unfolding scientific and technical logic. With this claim the company successfully fought a series of legal delaying actions, which gave it time to mobilize the messy and heterogeneous resources needed to generate a content and a context for success. In short, the *pretense* of a natural trajectory and the concealment of contingency behind legal and organizational barriers were central ploys in the process of *creating* a successful technology.

Bowker's story suggests that the *idea* that technology may be seen as the appliance of science is a powerful form of rhetoric but, at least in the case of Schlumberger, rather far from the truth. Bijker's chapter takes us to the very different history of the fluorescent lamp to make a similar point. Here the issue has to do with the relationship between invention, development, production, and diffusion. Bijker shows that the design of a high-intensity fluorescent lamp took place in what orthodox economic theory would call its diffusion stage. This lamp was not designed by engineers in research and development, but rather through the joint efforts of the executives of the electric light manufacturers and the utilities. In this case, then, the conference table became the drawing board!

Here again, heterogeneous economic, organizational, and technical contingencies were at work. When General Electric and Westinghouse launched their original version of the fluorescent lamp, they were clear that one of its attractions was its efficiency. But this meant that the new fluorescent lights might reduce the sales of power—a matter of deep concern to the utilities. The invention of the high-intensity fluorescent lamp met the concerns of both the manufacturers and the utilities.

These three chapters thus press home the message that technical change is contingent and heterogeneous. They also, however, show that it is possible to tackle the character of that change using a variety of different vocabularies and theoretical perspectives. Law and Callon make use of the actor-network approach, which rests on the idea that innovation and the strategies that shape it may be described in a network vocabulary that emphasizes the interrelated and heterogeneous character of all of its components, whether social or technical. It also puts forth the view that the social and the technical are established simultanously—indeed that they mutually constitute one another.

Bowker is also influenced by the actor-network approach, and in particular its concern with dealing evenhandedly with both the technical and its institutional context. However, his piece also draws on a range of other resources. In particular, his background as an historian is revealed in the analogy he draws between textual, contextual, and self-validating features of geophysical accounting and invention on the one hand, and the debate between Tawney and Trevor-Roper on the origins of the English revolution on the other. One consequence of this is the way in which he displays a concern with the products of historiography and the fact that they are ultimately open to question. Another is his interest in the way in which historical accounts may work to influence history and so generate the conditions for their own validity.

If Bowker brings the nuanced eye of the historian to his subject matter, Bijker's piece applies and extends a particular sociological tool to the analysis of technological change. The term "technological frame" refers to the concepts, techniques, and resources adopted by technologists and others. It is thus a way of talking of the set of theories, expertises, values, methods of testing, and physical tools and devices available to communities as they negotiate about the putative character of innovation. Here again, the stress is on heterogeneity. Bijker presses the view that both social groups *and* technologies are generated in the contingent arrangement of the concepts, techniques, and resources brought together in the relevant technological frames. Society itself is being built along with objects and artifacts.

1

The Life and Death of an Aircraft: A Network Analysis of Technical Change

John Law and Michel Callon

Imagine a technological project that lasts for a number of years, involves the mobilization of tens or hundreds of thousands of workers, designers, managers, and a plethora of heterogeneous bits and pieces including designs, parts, machine tools, and all the rest. Imagine that this project is developed in a constantly changing environment— that requirements, interests, and even the actors themselves change during the course of its lifetime. Imagine that not hundreds but hundreds of thousands of decisions are made. And imagine that in the end it is cancelled amid a welter of acrimony. How can we describe such a project in a way that is more than "simple" history? How can we describe it in a way relevant for the analysis of other projects and technological innovations? How can we explain the decision to close the project? How can we explain its failure? And how can we do this in a way that lets us avoid taking sides?

Despite the recent growth in interest in the social analysis of technology, few tools currently available are really useful. Our problem is that it is too simple (though it contains an element of truth) to say that context influences, and is simultaneously influenced by, content. What we require is a tool that makes it possible to describe and explain the coevolution of what are usually distinguished as sociotechnical context and sociotechnical content. In recent work we have used a network metaphor to try to understand this kind of process (Callon and Law 1989). We have considered the way in which an actor attempts to mobilize and stabilize what we call a *global network* in order to obtain resources with which to build a project. In our language, then, a global network is a set of relations between an actor and its neighbors on the one hand, and between those neighbors on the other. It is a network that is built up, deliberately or otherwise, and that generates a space, a period of time, and a set of resources in which innovation may take place. Within this space—we call it a *negotiation space*—the process of building a project may be treated as the elaboration of a *local network*—that is, the development of an array of the heterogeneous set of bits and pieces that is necessary to the successful production of any working device. We have suggested, that is, that the notions of context and content that are used as common analytical devices in the sociology of science and technology may be transcended if projects are treated as balancing acts in which heterogeneous elements from both "inside" and "outside" the project are juxtaposed.

In this chapter we push our analysis a stage further by considering the dynamics of a large British aerospace project. We consider the way in which the managers of that project sought to position their project in a global network in order to obtain the time and the resources needed to build and maintain a local network. And we discuss the way in which the shape of that project was influenced not only by the efforts of those managers, but also by events and strategies that influenced the shape of the global network. Thus we trace the strategies and contingencies that led to the creation of both local and global networks, the fortunes or the managers as they sought to shape both networks and control the relations between them, and the eventual collapse of the project when the relationship between them finally got completely out of hand.

At one level, then, our story is banal. It is the description of a large military technology project that went wrong. But although this project has considerable interest for the history of British aerospace, here our aim is not primarily to add to the catalog of accounts of military waste. Rather it is analytical. Like many others in this volume, we are concerned to develop a vocabulary of analysis that will allow us to describe and explain all attempts to build durable institutions. Analytically, the fact of the failure in the present project is best seen as a methodological convenience: controversy surrounding failure tends to reveal processes that are more easily hidden in the case of successful projects and institutions.

A Project and Its Neighbors

The TSR.2 project was dreamed up in the Operational Requirements Branch of the Royal Air Force (RAF) in the late 1950s. (TSR stands for Tactical Strike and Reconnaisance; the meaning of the 2 is a mystery.) The structure of the project and its aircraft were conceived in the course of a set of negotiations with neighboring actors. Thus, those who advanced the project sought to establish for it a shape that would allow it to survive. In some cases it was a question of securing sufficient resources from neighboring actors. In other cases it was a question of securing their neutrality for an appropriate period. In both cases it was a question of coming to appropriate arrangements—of defining the relationship between the project and its neighbors.[1]

The origin of this process can be traced to a General Operational Requirement (GOR 339) developed by the Operational Requirements Branch and to a policy for the rationalization of the aircraft industry implemented by the procurement branch of the British government, the Ministry of Supply. So far as the RAF in general was concerned, it was necessary that the end product be an aircraft. All other transactions were predicated on this assumption. That a combat aircraft was needed was not, in fact, that clear in the late 1950s. The defense policy of the United Kingdom as spelled out in the 1957 Defence White Paper was that of nuclear deterrence based on ballistic missile retaliation. So far as the Ministry of Defence was concerned, it was important that the end product not be a strategic bomber—this alternative having been ruled out by the White Paper. This suggested that the project should be a combat aircraft, and given British defense commitments as conceived by the Ministry, it was appropriate that it should be a tactical strike and reconnaissance aircraft (TSR).

So far as the Treasury was concerned, it was important that the end product be cheap. Given this perspective, which was based on its perceived need for economies in defense spending, the Treasury tended to doubt the need for any aircraft at all. At most support could be found for a single combat aircraft. This meant that the aircraft would have to fulfill all the possible combat aircraft requirements of the RAF. Accordingly, there was pressure for a versatile aircraft—a requirement fulfilled by the TSR definition—and also one that might be sold overseas, thereby cutting its unit cost.

So far as the Navy was concerned, it was also necessary to overcome a high degree of hostility. The Navy was purchasing a small tactical strike aircraft called the Buccaneer, and was anxious to persuade the RAF to buy this same aircraft because this would cut unit costs for the Navy and relieve pressure on the arms procurement budget overall. The response of the Operational Requirements Branch was to propose a large, supersonic, precision-strike, long-range aircraft that was quite different from the Buccaneer. Although this response was not what was sought by the Navy, it was intended to neutralize the (Treasury-assisted) attempts by the latter to impose the Buccaneer.

So far as the Ministry of Supply was concerned, it was important that the aircraft project be consistent with a policy for rationalizing the airframe and aeroengine industry. There were upward of a dozen airframe manufacturers in the United Kingdom in the late 1950s. The Ministry felt that there was room for two or three at most. Accordingly, the project was conceived as an instrument for bringing a large and powerful industrial consortium into being: it would not be awarded to a single firm.

These transactions shaped and helped to define the project. Let us note a number of important characteristics of this process.

The TSR.2 project displayed what we may call variable geometry: it represented different things to different actors. In other words, it possessed a high degree of "interpretive flexibility." For the Ministry of Defence and the RAF, it was not a strategic bomber but a tactical strike and reconnaissance aircraft. For the Treasury it was relatively (though insufficiently) cheap. For the Navy it was a successful competitor to the Buccaneer, and for the Ministry of Supply it was an instrument of industrial policy.

At the same time, however, it was also a relatively simple object to each of those other actors. Though our account is, of course, schematic, most of the complexities of the aircraft and its project were also invisible to these outside actors. But the simplification involved in bringing this project into being was reciprocal: the outside actors were, in turn, simplified from the standpoint of the project. Thus the Treasury was (and is) a highly complex bureaucracy with a wide range of policy concerns and procedures. From the standpoint of the project most of these were irrelevant. The Treasury was a "punctualized" actor—an actor that was reduced to a single function, that of the provision of funds.

This process of reciprocal simplification has several consequences. One is that from the standpoint of both its neighbors and an outside observer, the project can be treated as a series of transactions. Some of these took the form of economic exchanges: in return for the provision of funds the project would provide accounts, progress reports, and, ultimately, a working aircraft. Some were political in character: in return for a demonstrated need for a large and complex aircraft, the objections of the Navy to the project would be overruled. Yet others were defined technically (the General Operational Requirement, and the more specific Operational Requirement that followed it) or industrially (the provision of contracts in exchange for a rationalization of the aircraft industry). In an earlier paper (Callon and Law 1989) we referred to what is passed between an actor and its neighbors as *intermediaries,* and we will adopt this (deliberately general and nonspecific) terminology here to refer to what passes between actors in the course of relatively stable transactions. And, as indicated earlier, we will use the term *global network* to refer both to the set of relations between an actor and its neighbors, and to those between its neighbors.

It is also important to note that transactions leading to reciprocal simplification shaped not only the project itself but also the actors that entered into transactions with it. Again, this shaping operated through a variety of mechanisms: often the formulated *interests* of existing actors were redefined. In 1957 the Ministry of Defence did not "know" that it needed a TSR aircraft. It simply knew that it did not need a strategic bomber to replace the existing V bomber force because ballistic missiles would fulfill this role. In the process of interacting with the Operational Requirements Branch, the ministry was persuaded or became aware of its interest in a TSR aircraft. A similar process overtook the RAF. At the beginning of the process it knew only that it wanted a new combat aircraft, and that there were important obstacles to this ambition. By the end it perceived its interests in terms of the TSR.2. A similar but even more dramatic process overtook the airframe manufacturers. They started out with a general interest in obtaining contracts to produce new aircraft, and ended up finding that it was in their interest to merge with manufacturers that had previously been rivals to design and manufacture a TSR aircraft. So profound was the process in this case that they were not simply reshaped—they were turned into new actors in their own right.

However, the actors shaped by the project were not, in all cases, influenced by operating on their perceived interests. Thus the expressed interests of the Navy with respect to the project remained unchanged in the following years: it was hostile and wished to see it cancelled. However, because of the definition of the aircraft described above and a series of bureaucratic political ploys that will not be detailed here, the project and those whose support it enlisted (notably the RAF itself) boxed in the Navy. The latter was hostile, but it was also unable to press its hostility home. In this case power plays and bureaucratic strategems acted to shape the Navy. The neutrality of the Treasury was secured in part by similar means.

We are emphasizing this process of mutual shaping because it is important to understand that actors are not simply shaped by the networks in which they are located (although this is certainly true), but they also influence the actors with which they interact. In one way this is obvious, for the latter class of actors are themselves located in and shaped by a global network. However, the point is worth making explicitly because it breaks down an abstract distinction common in social analysis between (determined) actor and (determining) structure, or between content and context. Neighbors do indeed shape new actors as they enter into transactions with them, but they are in turn reshaped by their new circumstances.[2]

Finally, we should note that financial resources, a set of specifications, the tolerance of certain neighbors, and the neutralization of others offered the project managers the resources to go about fulfilling their side of the explicit and implicit bargains that they had entered into. In short, like many of the other cases described in this volume, the project had created for itself a time and a space within which it might deploy the resources it had borrowed from outside. It had, accordingly, achieved a degree of autonomy, a "negotiation space." We will now consider some of the transactions that took place within this negotiation space.

Designing a Local Network

By the autumn of 1957 the negotiation space for the project managers was quite limited. In general they were obliged to adopt a step-by-step approach: for instance, no funds would be forthcoming unless they produced intermediaries in the form of clearer ideas about the design of the aircraft, its likely manufacturers, the costs involved, and the probable delivery date. The first stage in this process was to specify the design features of the aircraft more fully. Thus GOR 339 was quite general, specifying the kind of performance required rather than detailing the design of an aircraft. The latter would be necessary if such skeptics as the Treasury were to be convinced that a consortium of manufacturers was indeed capable of producing the proposed aircraft within budget. Accordingly, the process of giving shape to the project continued. Now, however, the focus of the project managers turned inward: they started to try to elaborate a network of design teams, design features, schedules, and contractors. They started to create and mobilize actors in what we will call a local network.[3]

The first step in this process was to ask the British aircraft industry to submit outline designs in the autumn of 1957. This posed no particular problem, for the firms in question were hungry for work and readily mobilized. In all there were nine submissions (Gardner 1981, 25), though here we will mention only the three most relevant to our story (Williams, Gregory, and Simpson 1969). Vickers offered two possibilities. One was for a small single-engine aircraft that was relatively cheap but diverged considerably from GOR 339. The other was for a much larger aircraft that conformed closely to GOR 339. Both proposals advocated a "weapons systems" approach to design with an integrated approach to airframe, engines, equipment, and weapons (Wood 1975, 156). Although this represented a departure from traditional methods of military aircraft procurement in which airframes were designed, built, and tested first, and weapons and equipment were added afterward, the approach was well received in Whitehall, in part because of an extensive selling exercise by Vickers and in part because it accorded with Ministry of Supply thinking and recent American experience.

Nevertheless, although the general philosophy of the submission was clear, well articulated, and closely argued, Vickers were not able to do all the necessary design work and saw themselves going into partnership with another firm, English Electric, which had designed and manufactured the successful Canberra light bomber and the Lightning supersonic fighter. However, English Electric had made its own submission, code-named the P17A, which was a detailed aerodynamic and airframe design for a 60,000 to 70,000 lb. delta-winged Mach 2 strike bomber with twin engines and two seats (Hastings 1966, 30; Williams, Gregory, and Simpson 1969, 18; and Wood 1975, 155). Though the P17A met many of the specifications of GOR 339, it lacked an all-weather capability and a vertical or short takeoff capacity (Williams, Gregory, and Simpson 1969, 18). English Electric countered the latter deficiency by arguing that short takeoff was not the most urgent requirement (which was, in their view, the replacement of the Canberra), but suggested that this could be provided at a later date by a platform that would lift, launch, and recover the P17A in the air. This platform was to be designed and built by Short Brothers, which submitted a preliminary design (Hastings 1966, 29; Williams, Gregory, and Simpson 1969, 18; Wood 1975, 155).

With the airframe manufacturers mobilized and a set of submissions in place, the second stage in the elaboration of the local network started—consideration of what design or combination of designs would best fulfill the various requirements negotiated with neighboring actors. Though the small Vickers design was favored by the Treasury because it was likely to be relatively cheap, the large submission was particularly attractive to the Air Staff, the RAF, and sections of the Ministry of Defence. This was because it strengthened the commitment of the Air Staff both to a short-takeoff aircraft (which would have to be large because it would need two powerful engines) and to a weapon systems approach. The staff, the Ministry of Defence, and the Ministry of Supply were also impressed by the integrated design philosophy advocated by the company and were persuaded that Vickers had the management capacity to control and integrate a complex project (Wood 1975, 158; Gardner 1981, 33). However, they were also impressed by the English Electric submission, which was generally conceded to be "a first class design" (Wood 1975, 155), was the product of wide experience with supersonic aircraft, and also had the advantage that it could use existing avionics equipment in the short run. In addition, though contact between the two firms had been limited (with English Electric contractually tied to Short Brothers), Vickers had indicated its wish to have English Electric as its partner. Accordingly, the Air Staff came to the conclusion that a combination of the large Vickers-type 571 and the English Electric P17A would be both appropriate and capable of being used to mobilize actors in the global network.[4]

Accordingly, with a putative design and potential contractors in hand, the Air Staff returned to the global network in June 1958. Specifically, they went to the Defence Research Policy Committee (Gardner 1981, 32). This group was responsible for the overall control of defense procurement and as part of its role assessed and allocated priority to the projects put to it by user services and the appropriate supply departments (Williams, Gregory, and Simpson 1981, 32). Cabinet-level approval was ultimately obtained, and GOR 339 was replaced in early 1959 by a tighter, more technical and definitive requirement, Operational Requirement (OR) 343 (Gardner 1981, 33; Wood 1975, 158), and an associated Ministry of Supply specification, RB 192 (Gunston 1974, 41).[5] All was now in place: a preliminary network of local actors had been mobilized and had contributed to creating the intermediaries needed to satisfy the global actors or turn their objections aside. The design for a local network of firms, technical components, management procedures, and the rest had been approved. Intermediaries would start to flow from the global network in order to mobilize a more permanent local network.

The Creation of a Local Network

Vickers and English Electric did not wait for contracts to be awarded formally. In late 1958 they set about the difficult task of building a permanent local network of designers, designs, production teams, management, and subcontractors that would bring about the construction of a TSR.2 within the time and budget permitted by neighboring actors. The first step was to try to integrate and take control of two quite separate industrial organizations and designs. Several problems had to be overcome in this process of designing and mobilizing a local network. First, the designers who had previously worked in two teams some 200 miles apart had rather different approaches to design. Thus the Vickers team, which was based in Weybridge in Surrey and near Winchester in Hampshire, had concentrated on electronic systems, on airborne systems in general, on fuselage design and on short takeoff and landing (Williams, Gregory, and Simpson 1969, 29). The English Electric team was based on Warton in Lancashire and had concentrated on supersonic aspects of the design, the implications of low-level flight, and had, as we have noted, submitted the more detailed airframe design. The process of getting to know one another and settling down to collaborative work was difficult but generally successful in the end (Beamont 1968, 137; Beamont 1980, 134; Williams, Gregory, and Simpson 1969, 47), and a joint team of fifty designers undertook a detailed study of the technical and design problems raised by GOR 339 by the early months of 1959. Following this a division of labor evolved that reflected the relative skills of the two teams: the Weybridge group worked on systems including cost-effectiveness and weapons, while the Warton team worked on aerodynamics (Wood 1975, 164).

But the local network was not composed of people alone. For instance, the problems posed by the differences between the two designs were at first considerable. The most fundamental of these arose out of the different requirements suggested by supersonic flight and a short takeoff capability. High-speed flight suggested a small wing with low aspect ratio, a low thickness-to-chord ratio and a high leading edge sweep—all features of the P17A. A short-takeoff capability suggested the need for a low wing loading, which in turn implied that the wing should be large, and it also suggested a high thickness-to-chord ratio and a low leading edge and trailing edge sweep. Sir George Edwards, head of Vickers and later of the merged British Aircraft Corporation, is reported to have said at one stage, "The Vickers STOL study and the English Electric machine with a tiny low level wing . . . seemed irreconcilable" (Gunston 1974, 44). The team wrestled with these different requirements and eventually resolved them in a single solution by: a. providing very large flaps that increased both the thickness-to-chord ratio and the angle of attack; b. forcing high-pressure air over the flaps in order to improve lift at low speeds by preventing the breakup of airflow over the top surface of the wing; and c. increasing the thrust-to-weight ratio by specifying two extremely powerful engines (Gunston 1974, 46; Williams, Gregory, and Simpson 1969, 25, 39; Wood 1975, 165).

Although this was the most fundamental design decision—for given the Operational Requirement, many other decisions about engines, moving surfaces, undercarriage, and integral fuel tanks were seen by the team to be foreclosed—other and somewhat separable design difficulties also arose. One of these concerned the location of the engine. The necessity for thin, uncluttered wings suggested that these should be located within the fuselage, as in the English Electric design. Vickers were skeptical about this, worrying about cooling problems and the risk fire. However, in the end the English Electric view carried the day (Wood 1975, 163). Another concerned the short-takeoff capability of the aircraft. In 1959 the Air Staff were hoping for this, but the designers quickly concluded that the proposed aircraft was too heavy, and they sought—and were given— permission to build an aircraft that would take off instead from half runways and rough strips (Gunston 1974, 41).

In March 1960 the wing position was moved by three inches as a result of these and similar deliberations (Hastings 1966, 40; Gardner 1981, 105), but after this the design was changed little in concept, and a brochure and drawings were issued to the workshops in 1962 (Wood 1975, 165).[6] A putative local network of technical components had been specified. All that remained was to turn these from paper into metal.

Integrating their designs and their design teams were not the only problems of integration and control confronted by the two firms. There was also a question about how the production work should be allocated. Although the contract from the Ministry of Supply stated that the two firms were to share the work equally, it was also made clear that Vickers was the prime contractor and would exercise overall mangement control (Hastings 1966, 35; Williams, Gregory, and Simpson 1969, 22). This led to some ill feeling in English Electric, which felt that it should have received its own contract directly from the ministry. The problem was exacerbated by the commitment to a development batch approach. The prototypes and development aircraft would be built on the production line for the main series rather than being built by hand, separately. The location of the production line had, therefore, to be determined early on, and negotiations were difficult (Gardner 1981, 32).

Relations between Global and Local Networks

While the design and creation of a local network went ahead, there were continuing difficulties in the interaction between the local network and the global network that had brought it into being. As we have already indicated, in principle the Ministry of Supply was committed to a weapons systems approach to procurement—the whole machine including all its avionics, armaments, and other subsystems should be conceived as a whole. In the view of the Ministry, this approach had implications for management:

Since the failure of only one link could make a weapons system ineffective, the ideal would be that complete responsibility for co-ordinating the various components of the system should rest with one individual, the designer of the aircraft. (*Supply of Military Aircraft* 1955, 9)

The approach thus implied centralized control. It suggested that a single locus should shape and mobilize the local network *and* that this locus should have control over all transactions between the local and global networks. It should, in short, become an *obligatory point of passage* between the two networks.

As we have indicated, Vickers was indeed appointed prime contractor and was responsible in principle for controlling the entire project (Hastings 1966, 35; Williams, Gregory, and Simpson 1969, 22). In practice, however, the Ministry of Supply (later Aviation) did not vest all responsibility for control in Vickers. Rather, the project was controlled by a complex series of committees on which a range of different agencies were represented, and no single agency was in a position to control all aspects of the project. The failure of the management of the newly formed British Aircraft Corporation to impose itself as an obligatory point of passage led to a number of complaints by the latter about outside interference. These fell into two groups:

1. Actors in the global network were able to make (or veto) decisions that affected the structure of the local network:

a. Many of the most important contracts were awarded directly by the Ministry; the contract for the engines provides a case in point. The design team took the unanimous view that this should be awarded to Rolls Royce. This recommendation was based on the belief that a reheat version of the RB 142R offered the thrust-to-weight ratio necessary for the aircraft, was lighter, and had more potential than an alternative enhanced Olympus engine made by

Bristol Siddeley (Hastings 1966, 41; Wood 1975, 164). However, the Ministry of Supply had other views, apparently deriving from its concern to pursue an industrial policy of merger, and despite this recommendation awarded the contract to Bristol Siddeley (Clarke 1965, 77; Gardner 1981, 29; Gunston 1974, 41; Williams, Gregory, and Simpson 1969, 21). In fact, overall, the BAC controlled only about 30 percent of the project expenditure itself (Gunston 1974, 67; Hastings 1966, 40).

b. The Air Staff tended to make decisions without reference to the BAC. The problem here was that the RAF continued to develop its ideas about the ideal performance and capabilities of the TSR.2. This tendency to upgrade specifications was encouraged by the fact that contractors would often talk directly to the Air Staff and the Air Ministry. Sometimes such discussions would lead to changes in the specification of equipment whose specifications had already (or so the BAC thought) been fixed. One result was that, at least in the view of the BAC, progress toward freezing the design of the aircraft was impeded (Hastings 1966, 144; Gardner 1981, 101; Williams, Gregory, and Simpson 1969, 49).

2. Given the number of global actors that had a right to express their views in the committee structure, arriving at a clear decision was sometimes difficult.

a. It was often impossible to get a quick decision from the various government agencies. Hastings (1966, 160) describes the case of the navigational computer that was the responsibility of a firm called Elliott Brothers. The specification for this computer was very demanding, and Elliott concluded that the only way in which this could be met within the time allowed was by buying the basic computer from North American Autonetics. The Ministry resisted this because it had sponsored basic research on airborne digital computers in 1956–57. The Ministry ultimately accepted Elliott's view, but the equipment required was complex and the price was high. This brought into play Treasury representatives, who insisted that the decision be reviewed after a year. The whole argument delayed the development of the computer and (or so Hastings argues) added £750,000 to the cost.

b. On a number of occasions the Treasury used its position to try to cancel the project, or at least reduce its cost, and there seems little doubt that an initial delay in issuing contracts was in part a function of Treasury reluctance. When the committee structure was further elaborated in 1963, the opportunities for discussion about costs became greater still. Indeed, the Projects Review Committee, which included Treasury membership, had no representatives from industry (Hastings 1966, 38; Williams, Gregory, and Simpson 1969, 82).

c. The technical committees often made decisions with relatively little thought of cost, whereas those committees concerned with costs had little information about, or ability to determine, the technical necessity of the tasks they were examining (Hastings 1966, 35; Williams, Gregory, and Simpson 1969, 22). Certainly it appears that the RAF sought optimum performance in a way that was relatively cost-insensitive. (Hastings 1966, 59–60). The Air Staff tendency to delay was strengthened by the weapons systems philosophy and the development batch approach to procurement, both of which reinforced the RAF desire to be sure that the design was absolutely right before it was frozen, because it was so difficult to introduce modifications once this had occurred (Williams, Gregory, and Simpson 1969, 53).

Difficulties in Mobilizing a Local Network

We have described the reaction of the British Aircraft Corporation to the fact that outside actors refused to let it serve as an obligatory point of passage between the project's global and the local networks. However, the growth of mistrust between the Ministry and the BAC was two-way. The Ministry came to believe that the prime contractor was failing to exercise adequate management control (Hastings 1966, 157; Williams, Gregory, and Simpson 1969, 54). In particular, it was suggested that there was no single "iron man" at the BAC to direct the project (Wood 1975, 172), and at one point the ministry felt obliged to represent this view very strongly to the firm. Thus, although the Ministry's point of view has not been as well documented as that of the BAC, it is pretty clear that for much of the period after 1959 *neither* acted as an obligatory point of passage between local and global networks, and there was continual "seepage" as local actors lobbied their global counterparts, which influenced and in some cases impeded the smooth running of the project.

Indeed, the construction of the local network presented many problems. Perhaps the most serious of these concerned the engines. It is clear in retrospect that neither the Ministry nor Bristol Siddeley knew what they were letting themselves in for when the contract was awarded. The Ministry specified the engines in very general terms, and it was at first thought that their development would be a fairly straight-forward matter of upgrading an existing type, the Olympus (Williams, Gregory, and Simpson 1969, 27, 52). It turned out that this was not the case. The engine that was developed had a much greater thrust than its predecessor and operated at much higher temperatures and pressures. When it was first proved on the test bed, it turned out that its cast turbine blades were too brittle, and it was necessary to replace them with forged blades at considerable cost in both time and money (Hastings 1966, 42; Gardner 1981, 104).

This was not the only difficulty experienced by Bristol Siddeley. Serious problems arose with the reheat system, it proved impossible to install the completed engine in the fuselage, and there was also a weakness in the joint between the main engine and the jet pipe. However, the most serious problem appeared only late in the process of development. After proving the engine for over 400 hours on the test bed (Hastings 1966, 43), it was installed beneath a Vulcan in late 1962. On December 3 this aircraft was taxiing during ground tests at the BSE works at Filton in Bristol when the engine blew up, "depositing," as Wood (1975, 174) reports it, "a large portion of smouldering remains outside the windows of the company press office." The aircraft was reduced to burning wreckage, and although the crew was saved, a fire engine that approached the flames without due caution was caught up in the inferno (Gunston 1974, 56).

Within forty-eight hours it was clear that the failure had been caused by primary failure of the low-pressure compressor shaft. What was not clear, however, was what had caused this failure. Bristol Siddeley hypothesized that it might be due to stress and ordered that the thickness of the shaft be doubled. At the same time it ordered an exhaustive series of tests—a further, elaborately mobilized network of actors—to investigate the reasons for the failure. These led to further unpredictable and unexplained explosions. Finally, in the summer of 1964 the cause of the problem was diagnosed. In the original unmodified engine, the low-pressure shaft had turned on three bearings. However, the design team had become concerned that the middle of these three bearings might catch fire at the high operating temperatures; this bearing had therefore been removed and then, to provide the shaft with sufficient rigidity, the diameter of this shaft had been increased (Beamont 1968, 139; Hastings 1966, 43; Wood 1975, 174). Under certain unusual circumstances, the air between this shaft and its high-pressure neighbor started to vibrate at a frequency that corresponded to the natural frequency of resonance of the low-pressure shaft. When this happened, disintegration quickly followed. However, even with a diagnosis at hand, a solution was going to require further time and money.[7]

Not all of the local network problems concerned the engines. It also proved very difficult to control the subcontractors. As we have indicated, same subcontractors appealed over the head of the BAC to the ministry in order to obtain favorable decisions about costs (Hastings 1966, 36; Gardner 1981, 101). Others colluded with the air staff to specify equipment that was unduly sophisticated. Again, from 1959—and more so from 1962, when the political climate began to undermine the project—many subcontractors doubted whether the aircraft would actually fly. This feeling was a function of another kind of seepage between the local and global networks— specifically the knowledge that the project had powerful opponents in government. The subcontractors thus sought to protect themselves (and recover their costs in full within each contract) by charging high prices, and they also tended to give the work low priority (Beamont 1968, 143; Gardner 1981, 102; Williams, Gregory, and Simpson 1969, 28). In addition there was a tendency to charge a wide range of development work to the TSR.2 because it was the only advanced military aircraft project in Britain (Gunston 1974, 53; Gardner 1981, 102). In any case, much of the work was not amenable to precise costing in advance (Gunston 1914, 60; Williams, Gregory, and Smith 1969, 27, 51). Although the aim of the ministry and the BAC was to issue fixed price contracts as this became possible, this goal was not achieved for many of the most important areas of work because unanticipated technical problems arose or the specification of the equipment was altered.

The Global Network Reshaped

The consequences of the failure to build a satisfactory local network made themselves felt in a number of ways. The RAF had been promised that the TSR.2 would be available for squadron service by 1965, but it was clear, with the engines still unproved in the middle of 1964, that this deadline had substantially slipped. The Ministry of Defence had likewise been promised a vital weapon with which to fight a war in Europe or the Commonwealth by 1965. This was not going to be available. The Treasury had been promised a cheap and versatile aircraft. Though it is true that some of the blame for the cost overrun can be laid at the door of the Treasury itself, by 1963 the estimated cost of the aircraft had nearly doubled. The Navy, which had been hostile from the outset, saw the project swallowing up more and more of the procurement budget. By 1963, then, all the relevant actors in the global network, whether sympathetic to the project or not, saw it as being in deep trouble. It was simply failing to deliver the intermediaries to the global network that it had promised when it had been given the go-ahead. Thus, although the data in table 1.1 are calculated on a variety of bases and are not in all cases strictly comparable with one another, they sufficiently illustrate this general trend.

However, although these difficulties were serious, they did not necessarily mean that the project was doomed. If the necessary intermediaries could be obtained from the global network, it would be able to continue: funds from the Treasury, expertise and support from the RAF, political support from the Ministry of Defence, and specialist services from such departments as the Royal Aircraft Establishment—these would allow it to continue. The RAF and the Minister, though not necessarily the whole of the Ministry of Defence, remained strong supporters of the project. With the government committed, it was not possible for the Treasury, the Navy, or indeed, the hostile sections of the Ministry of Defence, to stop the project. Accordingly, the funds continued to flow. However, armed

Table 1.1
Estimated costs and delivery dates of TSR.2

Date of estimate	Development estimate	Production estimate	Total
January 1959	£25–50m	up to £200m	up to £250m
December 1959	£80–90m (for 9 aircraft)		
October 1960	£90m	c. £237m (for 158 aircraft)	c. £330m
March 1962	£137m		
January 1963	£175–200m		
November 1963			£400m (overall, Ministry of Aviation)
January 1964	£240–260m		
February 1964			£500m (overall, Ministry of Defence)
January 1965			£604m (overall, Ministry of Aviation) £670m (overall, contractors) (R&D and production of 150 aircraft)

with the knowledge that came from their participation in the cat's cradle of government and industry committees, the skeptics were in a strong position to undermine the project by indirect means. This involved taking the fight into a wider arena.

The project had been conceived and shaped within the context of a limited number of global actors. Government departments, the armed services, the aerospace industry—these were the relevant actors that had given life and shape to the project. Though sections of the specialist press had some knowledge of the project, public statements by ministers had been very limited, and until 1963 it had had a very low profile. Gradually, however, this started to change as new actors first learned about the project and then indicated their opposition to it.

The most important of these was the Labour Party, which had declared its opposition to "prestige projects" such as Concorde and TSR.2 and had promised to review them if it was returned to power in the next General Election. Labour views about the TSR.2 had been unimportant in the early days of the project, and indeed were unformed. However, by 1963 this was beginning to change. The Labour Party was riding high in the opinion polls, and a General Election was due by October of 1964 at the latest. Whispering in government and by other insiders and a series of admissions from the Ministries of Aviation and Defence about delays and escalating costs led the TSR.2 to became an object of political controversy from 1963 onward. This process was reinforced by a highly controversial set-back to the project—the failure to persuade the Australian government to purchase the TSR.2 for the Royal Australian Air Force. In a blaze of publicity, the Australians opted for the rival F111, an aircraft built to a similar specification by the American firm, General Dynamics.

Thus, although supervision of the project remained in Whitehall, the number of actors, including critics, involved in its surveillance multiplied in 1963. The cost of the project was officially given as £400m. in November 1963. However, the Labour Party Opposition argued that this was a gross underestimate and put the figure closer to £1,000m., an estimate that was fiercely disputed by the Government (*The Times*, Nov. 12, 1963, p. 5). Furthermore, the Opposition argued that cost was one of the major reasons for the failure to procure the Australian order, a charge angrily rejected by the Government, which claimed that the constant carping of critics in the United Kingdom had led the Australians to doubt whether the aircraft would ever be produced (*The Times*, Dec. 4, 1963, p. 7).

Other critics suggested that the aircraft had become too expensive for its role and too expensive to be risked in combat, *The Times* suggesting that at £10m. per machine, it was "the most expensive way yet devised of blowing up bridges" (Sept. 28, 1964 p. 10).

Further political disagreements centered around the role of the aircraft. The cancellation of the British ballistic missile Blue Streak in 1960, followed by the 1962 cancellation of the American Skybolt, which had replaced Blue Streak, had led certain commentators to speculate that it might be possible to use the TSR.2 in a strategic nuclear role. This suggestion (which had always been seen as a possibility within government) was picked up by the 1963 Defence White Paper (Omnd. 1936) and attracted criticism both from those who felt that the aircraft was neither fish nor fowl, such as *The Times* and *The Economist*, and the left wing of the Labour Party, which was committed to a policy of unilateral nuclear disarmament. Yet others including Denis Healey, the Labour defense spokesmen, concluded that this "strategic bonus" did not so much represent a change in the specification of the aircraft as an attempt by the government to persuade its backbenchers of the soundness of its nuclear defense policy (*The Times*, March 5, 1963 p. 14). Controversy also surrounded the continued delays in the first test flight. Healey highlighted the symbolic importance of the maiden flight when he claimed in Parliament at the beginning of 1964 that the BAC had "been given an order that it must get the TSR.2 off the ground before the election, and that (this) was a priority" (*The Times*, Jan. 17, 1964, p. 14). However, though he was much too professional a politician to let the Conservative government off lightly for its alleged incompetence, he was also much too agile to foreclose his own options by promising to cancel the project if the Labour Party were to win the General Election.

Endgame

By the autumn of 1964 the project was at a crucial stage. The local network was practically in place: the TSR.2 was almost ready for its maiden flight, albeit very much behind schedule and over budget. But the structure of the global network had altered. Disagreement was no longer confined to the Treasury and the Navy and the RAF, the Ministry of Defence, and the Ministry of Aviation. (Indeed, some of these agencies were starting to alter their views of the project.) The dispute was now public, and the Conservative Government had committed itself firmly and publicly to the TSR.2, while the Labour

Opposition, though reserving its position, was generally highly critical of the cost and utility of the project. The future of the project thus
depended on two factors. First, it was important to demonstrate the
technical competence of the project, and the best way to do this was
for it to have a successful first flight. This would reinforce the position
of those who wished to see the project through. At the same time, the
outcome of the General Election was also vital. Conservative success
would probably assure the future of the project. Labour victory
would call it into question.

The maiden flight took place just eighteen days before the General
Election. Roland Beamont, the test pilot, describes the rather sub-
dued group of engineers, technicians, managers, and RAF personnel
who assembled at Boscombe Down before the flight. Most knew, as
the large crowd beyond the perimeter wire did not, of the potentially lethal nature of the engine problem, and they knew that although its cause had been diagnosed, it had not yet been cured. In
fact the flight was highly successful, the aircraft handled well, and
there was no hint of the destructive resonance that had plagued the
engines. Deep in the election battle the Prime Minister, Sir Alec
Douglas Home, described it as "a splendid achievement" (Beamont
1968, 151). The aircraft was then grounded for several months in
order to modify the engines and tackle minor problems with the
undercarriage.

The General Election took place on October 15. The result was
close, and it was not until the following day that it became clear that
the Labour Party had been returned to power with a tiny majority
of five. The new administration started work in an atmosphere of
crisis as a result of a large balance of payments deficit, and it decided
to cap defence expenditure at £2,000 million. It also ordered a
detailed scrutiny of the various military aircraft projects and started
a review of the proper future shape and size of the aircraft industry
(Campbell 1983, 79). In February the new Prime Minister, Harold
Wilson, made it clear that the future of the TSR.2 would depend on
four factors: first, a technical assessment of the aircraft and its alternatives; second, the fact that although the overseas purchase of an
alternative aircraft would save £250 million, this would also involve
considerable dollar expenditure; third, the future shape of the aircraft industry, and the possible unemployment that would result
from carcelling the program; and fourth, the nature of the terms that
could be negotiated with the BAC.[8]

At the beginning of April spokespersons for the principal actors
in the newly reconstructed global network—the Cabinet Ministers

responsible for departments of government—met to take a decision. They considered three possible courses of action: to continue with the TSR.2; to cancel it and put nothing in its place; and to cancel it and replace it with the similar F111 (Crossman 1975, 191; Wilson 1971, 90). The Treasury remained hostile to the TSR.2 and accordingly sought cancellation. Although it was concerned that a large purchase of an alternative American aircraft such as the F111 would impose severe dollar costs, it was prepared to accept that an *option* for the purchase of this aircraft should be taken out on the understanding that this did not imply a firm commitment. The Ministry of Defence was also in favor of cancellation on cost grounds, and it was joined by those, such as the Navy, that favored the claims of other services and projects (Hastings 1966, 68, 70). The Minister of Defence was in favor of an F111 purchase, but there was same uncertainty whether Britain really needed this type of aircraft in view of the country's diminishing world role (Williams, Gregory, and Simpson 1969, 31). He was thus happy to take out an option on the American aircraft rather than placing a firm order.

The position of the Minister of Defence probably in part reflected a shift in the view of the Air Staff. The combination of delay and cost overrun, together with the much tougher policy of economies introduced by the new Minister of Defence, had convinced the Air Staff that it was most unlikely that there would be a full run of 150 TSR.2s, and this had led to doubt about whether it would be possible to risk such a small number of expensive aircraft in conventional warfare. For some officers this pointed to the desirability of acquiring larger numbers of cheaper aircraft that might be more flexibly deployed. In addition, though the technical problems of the TSR.2 appeared to be soluble, its delivery date was still at least three years away. Because the F111 was designed to essentially the same specification and was already in production, the RAF found this quite an attractive alternative (Reed and Williams 1971, 181).

The Ministry of Aviation was concerned that a decision to scrap the TSR.2 would seriously reduce the future capacity of the British aircraft industry to mount advanced military projects, and tended to favor cancellation, combined with the purchase of a lower-performance British substitute. However, most ministers, including the Minister of Aviation, believed that the industry was much too large for a medium-sized nation. The real problem was that there was not yet in place a policy about its future shape and size. Even so, the TSR.2 was costing about £1 million a week, and further delay in cancellation did not, on balance, seem justified.

In general, the government was concerned that cancellation would lead to unemployment. With a tiny Labour majority in Parliament, ministers were anxious not to court unneccessary unpopularity. Against this, however, ministers felt that the resultant unemployment would mostly be temporary: that many of those working on the TSR.2 would quickly be absorbed by other projects or firms.

Nevertheless, the decision was by no means clear-cut: there was no overall Cabinet majority for any of the three options (Wilson 1971, 90). A number of ministers—mainly, it seems, those who were not directly involved—wanted to postpone cancellation until a long-term defence policy was in place (Crossman 1975, 190). Overall, however, those who wanted to maintain the project were outnumbered by those in favor of cancellation with, or without, the F111 option, and the vagueness of the latter commitment ultimately made it possible for these two groups to sink their differences.

The cancellation was announced by the Chancellor of the Exchequer, James Callaghan, in his Budget Day speech on April 6, 1965. The result was political uproar as the Conservatives sought to voice their anger and frustration at what they regarded as a foolish and shortsighted decision. A censure motion was debated on April 13. Amid charge and countercharge, Minister of Aviation Roy Jenkins concluded the debate for the government by agreeing that the TSR.2 was a fine technical achievement:

But, to be a success, aircraft projects must be more than this. They must have controllable costs; they must fulfill the country's needs at a price that the country can afford; they must be broadly price competitive with comparable aircraft produced in other countries, and they must have the prospect of an overseas market commensurate with the resources tied up in their development. On all these four grounds I regret to say that the TSR.2 was not a prize project but a prize albatross. (*Hansard*, April 13, 1965, c.1283)

The result of the censure debate was a resounding victory for the Government: it secured a majority of twenty-six, and any residual Opposition hopes that the the project might, somehow, be saved were dashed when members of the small Liberal Party voted with the Government.

Conclusion

In this chapter we have shown that the success and shape of a project, the TSR.2, depended crucially on the creation of two networks and on the exchange of intermediaries between these networks. From the global network came a range of resources—finance, political support, technical specifications and, in some cases at least, a hostile neutrality. These resources were made available to the project and generated what we have called a negotiation space. This was a space and a time within which a local network might be built that would in turn generate a range of intermediaries—but most obviously a working aircraft—that might be passed back to the actors in the global network in return for their support. We have also noted, however, that there were continual seepages between the global and the local networks in the case of the TSR.2 project. Actors in the global network were able to interfere with the structure and shape of the local network, while those in the local network were able to go behind the back of the project management, and consult directly with actors in the global network. The result was that project management was unable to impose itself as an obligatory point of passage between the two networks, and the troubles that we have detailed followed.[9]

The history we have described offers further evidence for several important findings of the new sociology of technology. First, it illustrates the interpretive flexibility of objects—the way in which they mean different things to different social groups. Second, as is obvious, it represents a further example of the social shaping of technology—namely the way in which objects are shaped by their organizational circumstances (Pinch and Bijker 1987; MacKenzie and Wajcman 1985; Callon 1986; Law 1987; MacKenzie 1987; Mac-Kenzie and Spinardi 1988; Akrich, this volume; Bijker, this volume; Latour, this volume). Thus we have sketched out the way in which the TSR.2 aircraft changed in shape both literally and metaphorically during the course of its development, and the relationship between these changes and the compromises that grew up for a time between the relevant human and nonhuman actors—compromises that achieved, as we have seen, no final solidity but that were, in turn, reworked as a function of new circumstances in the local and global networks.

Thus back in 1957 what we might call *aircraft number one* did not have a physical shape at all in the minds of the Air Staff or the Ministry of Supply (see table 1.2). It was rather the performance specification—a role to be played—and some of the circumstances in which it should be built. And this role reflected their view of what would pass muster with other relevant actors. Thus, the RAF wanted a flying combat aircraft, but the Ministry of Defence had a view of the future that left room for neither a strategic bomber nor a fighter.

Table 1.2
Three aircraft

Aircraft shape	Interested actors (+ definition of aircraft)	Hostile actors (+ definition of aircraft)	Neutral actors
1 • long range • supersonic • low altitude • STOL • all weather • large	RAF: • combat aircraft • in and out of Europe • dispersable • precision bombing/reconnaissance Defence: not strategic bomber	Navy: • Buccaneer Treasury: • cheap, versatile Buccaneer?	Labour party • in ignorance
2 • wing shape, delta, thin • two powerful engines • blown flaps • engines in fuselage • twin engines • integral fuel tanks	RAF: • large, twin-engine, sophisticated • TSR aircraft • STOL • long range Defence: • TSR aircraft BAC: • STOL difficult • VTOL impossible	Navy: • (blocked) Treasury: • (blocked)	
3 • option on F111 • TSR.2 cancelled	BAC: • buy 140 Conservative party: • TSR.2 essential Unions: • maintain work	RAF: • buy cheaper, more certain aircraft Defence: • buy cheaper aircraft Treasury: • cap expenditure • limit overseas spending Navy: • adopt Buccaneer Aviation: • buy chepaer U.K. aircraft Labour party: • cancel	

A tactical bomber and reconnaissance aircraft was the only remaining possibility—an aircraft that would play out specific, nonstrategic roles in Europe and British dependencies overseas. By contrast, the Treasury was quite uninterested in the defence of the Western Alliance. Much more important was the defence of the public purse in the face of ever more costly military technologies. Accordingly, it wanted no aircraft, or (second best) an existing aircraft, or if this was not possible (third, fallback, option), then no more than *one* type of new aircraft. The RAF judged it could force the Treasury to its fallback position, so it responded by specifying a single versatile aircraft. The Navy had strong views about defence needs, but it saw these in its own, quite different, carrier-based way. Accordingly, it wanted the RAF to procure a version of its small, subsonic Buccaneer. In a more negative sense, this was a strong incentive for the RAF to argue the need for a large, supersonic aircraft that was qualitatively different from its naval rival. And the Ministry of Supply wanted an aircraft that would be built by a consortium of firms rather than one alone.

Though it was touch and go, the Air Staff judged things rightly and the global network required by this shadow aircraft number one was stabilized. The result was *aircraft number two*—this time one that had, albeit on paper, a physical shape. This shape was partly a function of the global network of institutional actors mentioned above. But many other actors, considerations, and negotiations helped to structure the design. Thus the shape of the wings represented a compromise between the demanding specification required by the RAF on the one hand, and design skills, knowledge of aerodynamics and materials strengths, and the practice of wind-tunnel testing on the other. How on earth was short takeoff and landing to be reconciled with high-altitude Mach 2.5 flight and low-altitude, low-gust response? The wing was the physical answer to this question. It represented a compromise between these different considerations. But it also represented a compromise between the English Electric and Vickers design teams—in which English Electric had the upper hand. Similar reasoning—again in favor of English Electric—led to a decision about the location of the engines. These, it was decided, would lie within the fuselage to clear wing surfaces and avoid undue differential propulsive force in case of single engine failure—and this despite the potential fire hazard that so concerned the Vickers team. And it is possible to travel through the aircraft explaining the shape of each system as a physical compromise between the specification, the design teams, and a range of inputs from aerodynamics to the views of experts at the Royal Aircraft Establishment.

It can be argued that aircraft number two grew out of aircraft number one. Certainly many of the constraints and resources that went to shape number one helped to shape number two. But the process is not one of unilinear development. Aircraft number two was not simply the "unpacking" of a set of implications that were built into aircraft number one. Aircraft number one posed a set of problems to which there were many possible solutions. Aircraft number two represented a particular set of solutions to those problems—compromises negotiated by further numerous actors. Or, in some cases at least, it represented refusal to accept the problems posed by GOR 339, as is most obvious in the case of the short takeoff and landing requirement where the available rules of aerofoil behavior overruled the wishes of the Air Staff. In this instance, then, we see (if anything) the obverse of the social shaping of technology: it was the technical around which the social was being bent.

But if aircraft number two represents a translation rather than a simple development of aircraft number one, a translation shaped by a set of compromises between a somewhat different set of actors, then the metamorphosis of the project is yet more obvious for *aircraft number three*. This, which is more usually known as the F111, gradually took shape after the General Election. Thus we have traced the changes that took place among many of the most important actors after October 1964. The Treasury imposed rigorous economies and expressed extreme concern about the ever-increasing costs of the TSR.2 project, its short run, and its lack of export prospects. The Ministry of Aviation sought to shape a smaller and better-adapted aircraft industry. The Ministry of Defence was involved not only in cost cutting but also in a Defence Review that might lead to the abandonment of many British overseas responsibilities and with it, part of the rationale for the TSR.2. The Air Staff were increasingly concerned that they would not obtain the full 140 TSR.2s. For their different reasons *all* of these were prepared, with greater or lesser enthusiasm, to abandon the TSR.2 and take out an option on the F111. Accordingly, the project for a tactical strike and reconnaissance aircraft had been reshaped yet again by the relations between the actors involved, and with that reshaping the object that lay at its focal point had undergone metamorphosis yet again. This reshaping is summarized in table 1.2.

So much for the shaping and reshaping of TSR.2.[10] But how should we describe such a "translation trajectory?"[11] This, then, is our third concern. If technologies are interpretively flexible, if they are shaped by their contexts but they also shape the latter, then can we say nothing general about the contingent and iterative processes that generate them? Our answer, as we hinted in the introduction, is to deploy a network vocabulary and, specifically, to make use of the concepts of *global network*, *local network*, and *obligatory point of passage*. Our proposal is that the shape and fate of technological projects is a function of three interrelated factors.

The *first* is the capacity of the project to build and maintain a global network that will for a time provide resources of various kinds in the expectation of an ultimate return. Note that the successful construction of a global network has a specific and important consequence: it offers a degree of privacy for project builders to make their mistakes in private, and without interference—it offers a negotiation space (see Callon and Law 1989). In the ideal case the project builder thus obtains a degree of autonomy in its attempts to generate a return. It also—again in the ideal case—achieves both complete control over and responsibility for those attempts.

The *second* is the ability of the project to build a local network using the resources provided by the global network to ultimately offer a material, economic, cultural, or symbolic return to actors lodged in the global network. Put less formally, it is the ability to experiment, to try things out, and to put them together successfully. It is also the ability to control whatever has been produced and feed it back into and so satisfy the understandings that have been entered into with other actors in the global network.

The *third* factor, which is entailed in the first two, is the capacity of the project to impose itself as an obligatory point of passage between the two networks. Unless it is able to do so, it has 1. no control over the use of global resources that may, as a result, be misused or withdrawn, and 2. it is unable to claim responsibility in the global network for any successes that are actually achieved in the local network. It is, in short, in no position to profit from the local network.

Note, now, that the objects and actors in *both* global and local networks are heterogeneous. Thus in the case of the TSR.2 we mentioned a range of important institutional actors in the form of Whitehall ministries. But we also touched upon geopolitical factors (the presumed interests of a range of nation states) and technological changes (the advance of missile and anti-aircraft technologies). And we might equally well have considered the role of such naturally occurring features as prevailing winds (they were vital in the calculation of ferry ranges), and terrain cross-sections (which went into the calculation of the risks involved in low-level flying), or, for that matter, such human geographical but global considerations as the availablility and distribution of airstrips of different lengths.

But if global networks are heterogeneous, then so too are local networks. The TSR.2 project mobilized institutional actors in the form of contractors, subcontractors, and specialist agencies such as the Institute for Aviation Medicine. It mobilized tens of thousands of draftsmen, designers, market personnel, and fitters. It involved the use of a great body of high-status knowledge in the form of scientific and technical expertise and a large amount of equally important shop-floor knowledge and skills. And it involved numberless machine tools, jigs, motor vehicles, chaser aircraft, and test rigs, not to mention an awesome quantity of paperwork in the form of drawings, instructions, management charts, brochures, sales pamphlets, maps, and publicity handouts.

If the elements that make up global and local networks are heterogeneous, then the extent upon which they can be depended is also problematic: the degree to which they may be mobilized is variable, reversible, and in the last instance can only be determined empirically. In other words, the extent to which it is possible for a project to control its two networks and the way in which they relate is problematic, and it is the degree and form of mobilization of the two networks and the way in which they are connected that determines both the trajectory and success of a project (figure 1.1).

Concentrating on the two networks, it is possible to plot any project in a two-dimensional graph, where the x axis measures the degree of mobilization of local actors (control over local network) and the y axis measures the extent to which external actors are linked (control over global network). Furthermore, it is possible to describe the translation trajectory of any project (figure 1.2).

Thus, in the case of the TSR.2, the project started in the center of the diagram and climbed up the vertical axis as it sought to distinguish its product from the Buccaneer (A). Then, as the management structures were elaborated, it sought to move along the x axis to the right (B), and this tendency was strengthened as a design was agreed between the two former design teams, which in turn facilitated the formation of a single, unified design team (C). However, this position was not maintained. Little by little, as the subcontractors failed to fall into line, and in some cases interacted directly with the RAF, the degree to which the project management monopolized the internal network declined (D). This process reached a nadir when the low-

Local network   Global network

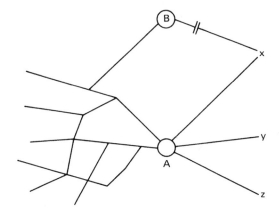

Strong external attachment
Strong internal mobilization
Strong obligatory point of passage

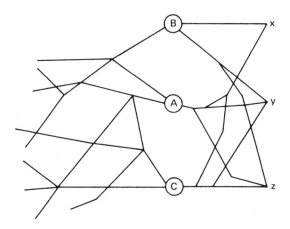

Weak external attachment
Weak internal mobilization
Weak obligatory point of passage

Figure 1.1
Strongly and weakly mobilized networks.

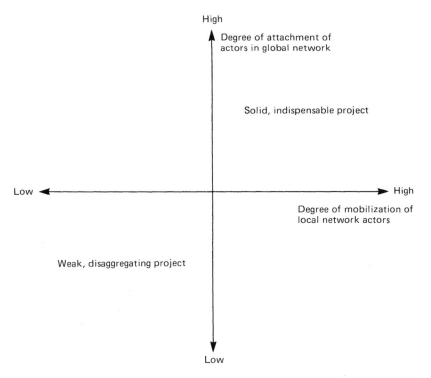

High

Degree of attachment of
actors in global network

Solid, indispensable project

Low   High

Degree of mobilization of
local network actors

Weak, disaggregating project

Low

Figure 1.2
Mobilization of local and global networks.

pressure shaft of the engine disintegrated and the latter blew up (E), and the Australians opted to purchase the F111 (F). However, after much remedial work the successful maiden flight took place and a degree of control over the local network was reasserted (G). Accordingly, the project moved back into quadrant 1, but with changing political circumstances and the availability of the F111, it reentered this quadrant lower down the y axis. Finally, with the election of a Labour government, the F111 came to be seen as a realistic alternative, and the project slipped down into quadrant 4 (H), and with cancellation it concluded by losing complete control of the local network, so ending up at the lowest point in quadrant 3 (I) (see figure 1.3). The major turning points in the trajectory of the project across this diagram can be depicted as a table of choices and consequences (see table 1.3).

We conclude, then, with the thought that the trajectories of technological projects are contingent and iterative. Sometimes, to be sure, a project or a technology may move forward in a manner that

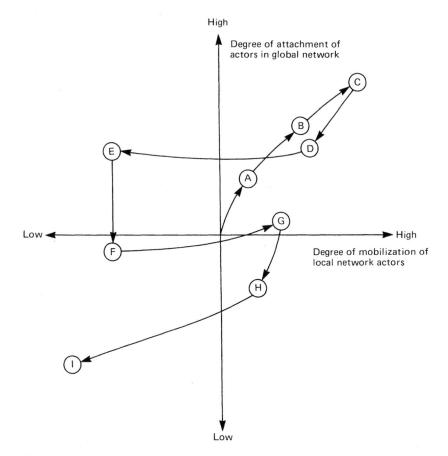

Figure 1.3
The trajectory of TSR.2.

accords to the stereotypical representation of the process of research and development. There is, however, no *necessity* about such progress. If all is smooth, this is because contingency has operated in that way. The kind of erratic progress we have described is far more likely— though such contingencies are often concealed in the Whiggish histories that celebrate the necessity of the successful after the event (see Bowker, this volume).

But our object is to move beyond the claim that outcomes are the product of contingency. Though this is right, it is also unhelpful unless we are content to accumulate specific case studies. Our aim is rather to seek patterns in the case studies. We believe that the case of the TSR.2—like a number of others in this volume—suggests that a crucial strategic move in building many, perhaps all, obdurate sociotechnologies is to create a distinction between inside and out-

Table 1.3
Choices and consequences

	Events/decisions	Local consequences	Global consequences
A	To build a new aircraft	Articulate design	Navy and Treasury blocked
B	Appointment of prime contractor	Articulate weapons	Minimize outside intervention
C	Decision about design	Develop production facilities	Secure funding
D	Support prime contractor's choices	Undermine prime contractor	Permit direct RAF intervention
E	Destruction of engines	Delay, mobilization of new teams and facilities	Expense and increased skepticism
F	Australian purchasing decision	Increasing skepticism by subcontractors	Increased politicization of project
G	Maiden flight	Technical confidence in aircraft and contractor	Strengthens supporters of project
H	Labour party wins election	Increases doubts among subcontractors	Strengthens opponents of project
I	Cancellation	Dissolution of project	Option to purchase F111

side, between backstage and front stage. The methods and materials for building such backstage negotiation spaces and relating them to the front stage are varied, and as the case of the TSR.2 shows, they are certainly not a function of strategy alone. We make use of a network metaphor because we need a neutral way of talking about the barriers that shape, for a time, the seamless web of sociotechnology.

Notes

John Law gratefully acknowledges the award by the Nuffield Foundation of a Social Science Research Fellowship, which made possible the empirical research on which this paper is based.

1. Here we adopt the methodological adage of Latour (1987) and "follow the actors."

2. In an earlier paper (Callon and Law 1989) in which we developed this argument in greater detail, we referred to these neighbors as "preforming networks."

3. Fuller details of this process of design are reported in Law 1987.

4. Little is known about the actual process by which decisions were reached. The best information available to us amounts to little more than hints. It does appear,

however, that the Treasury and the Ministry of Defence were fought off again in February 1958 (Wood 1975, 158). The Treasury was still concerned about the cost of the whole project, and the Ministry of Defence, noting the smaller of the two submissions from Vickers, toyed with the idea of specifying an aircraft that would fulfill some GOR 339 requirements and also be capable of carrier-borne operations (Wood 1975, 156). However, the RAF's need for a large aircraft of the TSR type was pressed both formally and informally, and GOR 339 emerged unscathed.

5. This specified that the TSR.2, as it was coming to be known, should be capable of high-altitude supersonic flight and a 1,000-nautical-mile radius of operations in a mixed sub- and supersonic sortie. It should also be capable of low-altitude treetop-level flight, have a terrain-following radar, display a low gust response, and have a short takeoff capacity, which in turn entailed a high thrust-to-weight ratio. It should have precision, self-contained navigational aids, be capable of delivering both nuclear and high-explosive bombs, have advanced photographic and linescan capabilities, and be reliable in order to minimize losses and permit operation from poorly equipped forward bases. Finally, it should have a ferry range of 3,000 nautical miles and be capable of inflight refueling.

6. In its definitive form the proposed aircraft had 1. a cruising speed Mach 0.9–1.1 at sea level and Mach 2.05 at high altitude; 2. a sortie radius of 1,000 nautical miles, 3. a takeoff capability of 3,000–4,500 feet on rough surfaces; 4. a climbing rate of 50,000 feet per minute at sea level; 5. a takeoff weight of 95,000 pounds for a 1,000-nautical-mile mission; 6. a high-wing delta configuration with large blown flaps but no control surfaces; 7. a large tailplane with all-moving vertical and horizontal surfaces; 8. two internally mounted Olympus 22R engines; 9. an internal weapons bay; and 10. an internal fuel capacity of 5,588 gallons.

7. The development of the engine and the detective work involved in diagnosing the cause of its failure is discussed in detail in Law 1992.

8. In January the government considered an offer from the BAC to manufacture 110 aircraft at a price of £575 million, with the firm picking up the first £9 million of any cost overrun (*Flight International* 87, 2928, April 22, 1965, p. 622). It did not accept this offer primarily because it was not prepared to carry all additional losses.

9. The limits to organizational power are usefully discussed in Clegg 1989.

10. Although it is outside this story, the aircraft went through a further reshaping in 1967 when the F111 was canceled. At that point aircraft number 4—a further version of the Buccaneer—entered the scene.

11. The notion of "translation trajectory" is, of course, ironic. Translations are the product of continual negotiation. They are precisely not the result of momentum imparted at their point of origin. We use the term to indicate the way in which our concerns overlap those of trajectory theorists—see, for instance, Sahal 1981, Dosi 1982, and Nelson and Winter 1982—but offer an analysis of technical change that is quite different in kind.

2

What's in a Patent?

Geof Bowker

In this essay, I am concerned with the kinds of accounts given of technical objects in patents, scientific literature, and company archives and in the relationships among the differing presentations of patents in these various sources. Numerous authors have pointed to the importance of patents in industrial science. In a notable turn of phrase, David Noble asserted that "Patents petrified the process of science, and the frozen fragments of genius became weapons in the armories of science-based industry."[1] Thomas Hughes (1983) has highlighted the fact that the research laboratory at General Electric was set up on the advice of the patent lawyer;[2] Reich (1985) has shown that in the Bell Company, industrial research was encouraged only when the strategy failed of buying up patents, then defending them in court.[3] Dennis and Bowker have identified the to-and-fro between patent lawyer and industrial laboratory as a key feature of industrial science.[4] In a ground-breaking essay, Cambrosio, Keating, and Mackenzie (1988, forthcoming) discussed the parallel between sociological and legal discourse about inventions and concluded that lawyers attacking patents draw on the same repertoire of analytical tools as the externalist sociologist.[5] I intend to develop this new perspective by looking at the ways in which patents are defended both in the courtroom and in the field. I will draw on the example of one company, Schlumberger, to discuss the relationship between the official version of history written into the patent and the actual use made of the patent.

In looking at patents as texts, I will concentrate on two features common to them all: they give internalist and Whig accounts of the development of the process or apparatus that they describe, and as legal instruments they attempt to impose that interpretation on the material world.[6] Now that within the history of science the ramparts of internalism and Whiggism have transmuted from stone to straw, *we* know that any account couched in these terms is necessarily false.

Further, all actors in the situations we describe (with the possible exception of the trial judge) themselves concede (outside of the courtroom) that such an account is false. And yet an immense amount of articulation work is done in an attempt to create a situation in which it can stand up in a court of law. We will be looking at this articulation work, and in particular at the question of who knows what about the patents and how this knowledge affects the success of the companies involved. To borrow a phrase from the sociological literature, we will be looking at the various awareness contexts within which patents function and at the articulation of the relationships among these contexts.[7]

To structure the analysis, I will draw an analogy between the process of defending a patent in the courtroom and that of defending a position within the discipline of history. The analogy is a natural one: these are among the only fields in which an "authorized" version of events (an historical occurrence or a scientific/technological discovery) is produced by the discussion of documents written to fit strict formal codes. The purpose of the comparison is to throw into relief three forms of relationship between "what actually happened" and what gets written about it. These three forms are: (i) the narrative's internal content and its immediate validity; (ii) the narrative's institutional setting, and how this reflects back on its truth; and (iii) the contribution of the narrative to making itself true. To illustrate my point, I will take the debate between Richard Tawney and Hugh Trevor-Roper about the origins of the English revolution.[8]

This debate provides a clear example of a structural framework that can, I believe, be fruitfully applied to the analysis of the defense of patents. The English revolution in 1640 saw the usurpation of the king's power by Parliament, in the form of Oliver Cromwell, and was the occasion of a famous series of debates (the Putney debates) about the nature of personal and religious liberty. Tawney gave a fairly classical Marxist account of this event: for him it was spearheaded by members of the "rising middle class" who were for the first time flexing their political muscles, using the language of universal freedom to pursue their own interests (as would the French revolutionaries later). For Trevor-Roper, the revolution was led by the "declining gentry," who were frightened of the increasing power of the bourgeoisie and took power in a desperate attempt to salvage their own privileges. Already, then, there is a clear parallel with a patent battle: two diametrically opposed accounts of a historical process clash in a public forum—academic journals in the one instance and the courtroom in the other.

The structural parallel that I want to develop relates to the three levels at which the debate played itself out: in journals, in institutions, and in the world at large. The first level is the internal one of the debate itself, involving the exchange of arguments in journals according to fairly strict rules of academic behavior. Archives were quarried for details of estate management and income; genealogies were drawn up to chart membership of the middle class and the nobility.[9] The equivalent level for the technical object is most clearly the courtroom battle about patents and the attached legal research, including the induction of expert witnesses. We shall see that here too competing histories were at stake and were being defended within a fairly strict framework. Most notably, both for the debate and for the patent battle there is an explicit belief that this cloistered, rule-driven activity can decide a historical truth. In both cases, many of the actors concerned recognize explicitly, outside of the academic presentation or the courtroom, that the debate is really decided elsewhere; and yet in both cases there is a vested interest on the part of actors in protesting that the show is all. This is the central aporia of this essay.

The second level at which the historical debate was played out was the institutional one. "Schools" were formed; journals favored one side or the other; a "Trevor-Roper" candidate could not get a job in a "Tawney" department. At this level, the appropriate weapons were not the telling argument or the subtle "mot" but the manipulation of research grants and access to publications—in short, occupation of the academic terrain. It was generally important that one's own side was bearing up respectably in the internal debate, but this debate was no longer crucial. The analogy with the patents process here is also clear. The company strategy is to occupy the terrain; and patents are a part of this process, but only one part. The patent must not only win in the courtroom, but the technical object attached to it must also impose itself in the real world. Control of the infrastructure was at stake. There is a figure/ground problem here. For those in the courtroom, the truth of the historical account was all, and the work of creating the infrastructure was background. For those outside, the historical account of the patent was in turn irrelevant background noise to the *real* focus of company activity. Strategies for imposing patents outside of the courtroom involve all kinds of different uses for them and sometimes ways to work around them. We shall see how companies could skirmish at this level and maintain a discourse about patents that denied this skirmishing.

And there is a third level at which the Tawney/Trevor-Roper debate played out: that of history itself. Tawney's position was inserted in a world in which there was continual progress toward socialism, Trevor-Roper's in a world in which socialism could never be other than a chimera. Indeed, each was trying through their writing to contribute to these respective outcomes. Their debate was interventionist in the same way that patent battles are: if the story is told well enough, then there is a chance that it will become true before the ultimate tribunal—history itself. If there were another revolution in England, a classic proletarian one this time, then Tawney would come out institutionally triumphant whether or not he was winning in the journals or in academia; if socialism crumbled there, then Trevor-Roper would win the day. One position would win out, and the fact that it had won would make the historical account it gave of seventeenth-century England true—at least as far as later historians were concerned. To put it simply, the fact that Marxism appears to be dead in Britain is a problem for Tawney's position on the English revolution.

The analogy with patent here is that the patents qua official history only survive within a certain framework of the state of the art and the state of the industry. If they *survive* the institutional and court battles, they might either prove to have described how the industry was shaped or prove to have been outside of the mainstream. The historical debate and the patent battle both seek simultaneously to describe a past reality (and impose that description) and to create a present one (and impose that creation). This is another form of aporia, because the name of the game in the debate is for the right hand (the intellectual one bearing rapier) not to know what the left hand (bearing club) is doing.[10] Success in debate and sufficient control of the infrastructure while the world changes enough to make you right are both vital; but there are no public forums at which historians writing histories or companies producing technical innovations admit to attempting both simultaneously.

There are two works within the field of the history of science and technology that seek to deal with this kind of broad sweep of social change. The first is Michel Callon's highly original article on electrical cars,[11] which gives a clear instance of this type of implication of the technical object in the process of social creation (and the creation of sociological theory). The other is Shapin and Schaffer's work on the air pump, where at stake in the development of scientific protocol is the imposition of a social form wherein the science of oneself and one's opponent would become respectively true and false.[12]

We now have a focus and a structure, so I will turn to a presentation of the final player—the patents and their attached technical objects. I have described in detail elsewhere how Schlumberger's electrical logging operations worked;[13] here I will give a brief overview, which should be sufficient to our purposes. Schlumberger started out in the business of searching for deposits of metals and oil in the period after World War I. Their technology was electrical. Their canonical technique was to run a current between two point on the surface of the earth, chart the electrical field that was generated, and use a variety of mathematical tools and theoretical graphs to interpret that chart. A development of this approach furnished their second main tool: in this instance they did not generate a field but charted variations in the earth's own electrical field. These techniques met with limited success worldwide, being overshadowed quickly by the much more accurate seismic methods. Their use did serve to bring Schlumberger to the attention of the oil companies, though.

The breakthrough for Schlumberger came in 1927, when it was decided to try exactly the same electrical techniques *inside* well holes as they were being dug. Now, rather than having to tease information out of approximate data covering vast terrains, Schlumberger could deal with small electrical variations that operated over a few feet. Further, the competitor was no longer the seismic method (which at that stage was not finely enough tuned to operate in the well hole) but mechanical coring (sample taking), which was notoriously inefficient, slow, and expensive.

The method was incredibly successful in the search for oil. Schlumberger's first measurement served essentially to distinguish strata of high and low resistivity. Oil was particularly resistive, water layers were conductive. It served to *correlate* fields (charting the continuations of a known field in accord with geological data) and to indicate the water shut-off point (the lowest point to which you could drill without hitting water—clearly valuable information for maximizing a field's production). It was not enough, however, to indicate the presence of oil or gas. The problem was that granite was just as resistive as oil. The analogy of the measurement of the earth's own field made the difference, because spontaneous activity in the earth's field was high in porous layers and low in nonporous ones. The only resistive porous layers, so the story went, were oil-bearing. Of course this overview is, as we shall see, itself a fabrication. It is, however, a neat and easy one that gives some view of the official history of Schlumberger.

With everything in place, I will rephrase the question motivating this essay. I have asserted that patents necessarily proffer a Whig interpretation of history, and an internalist one. I assume that life (let alone oil) is not really like that and so ask: how is the space created such that these interpretations can, locally and temporarily, hold? Why is this space created? This question comes down to one of the central problems of this book: how and why is the boundary between "inside" and "outside" created in scientific work?

Patents and Publications

One thing that stands out from the trial is that Schlumberger tried at all costs to impose an internalist view of events, whereas its opponent, Halliburton, tried to impose an externalist account.[15] For Halliburton, Schlumberger's patents did not describe the technical object used in the field, which had been modified to meet local conditions. Further, the logs that were produced were not universal but specific to a time and a place. There was no inner truth attached to the patent; environment was all. To prove this, they demonstrated that a Schlumberger logger could not interpret a given log without knowing where it came from. Thus for Halliburton, Schlumberger's patents put a universalistic gloss on a local and limited method. Not content with denying the present status of the technical object described by the patent, Halliburton tried to insert it into a different history. They went back to the 1830s to find analogues of the method in the Cornish tin mines. They argued that there was no rupture between this method and Schlumberger's own.

To formalize the differences between the two presentations, we can consider them along two axes. The first is that of rupture/continuity. Halliburton claimed that there was a rupture *now* between their methods and Schlumberger's (because what they are measuring is totally different); but that there was none *then* (between the past and the present). Inversely, Schlumberger claimed that there was a rupture then (marked by the original act of invention) but there is none *now* (when Halliburton is infringing on their patents). The second axis is that of the internal/external division. For Halliburton external factors apply *now*, because the patents conceal the actual, local truth of Schlumberger operations. For Schlumberger external factors applied *then*, before in a stroke of pure science Conrad Schlumberger formalized and rendered scientific the electrical treatment of the subsoil. This formal grid gives us a good idea of the framework of the discussion. As in all foundational stories of new science, the externalist explanation works for everyone before the founder.[16] The grid only makes sense if both parties assume that there is positive value in a new method being distinct from past forms and an improvement on them, and being governed by merely internal factors. At stake in the trial would be the allocation of points of rupture (then or now) and of significance to the local (then or now): the judge would decide the "correct" Whig, internalist account of the process of invention.

Schlumberger's whole publication strategy was geared toward the production of this correct account. Let us begin by looking at the clear pressure on the company to publish something about their methods. The first pressure came from the very newness of the technique itself. In a textbook in 1940, Heiland noted that "Progress and development in most geophysical methods have been largely the result of preceding developments in geophysical science. In gravitational, magnetic and seismic methods field procedure and methods of observation are closely allied to those in pure geophysics. Electrical methods lacking this background have followed their own course of development."[17] Not only was there no well-developed theory that Schlumberger could wave a hand at, but the whole area was under suspicion: "The quack and the shyster seem to have a strong predilection for electrical vestments. Another unfortunate circumstance is that electrical trappings are, in the minds of many laymen, endowed with mystical power."[18] Gish's statement is amply borne out by developments both at the time (attempts to measure oil reserves in Russia from measurements of the electrical field in a laboratory in Paris)[19] and since (a recent electrical hoax proved expensive for the French government). So Schlumberger had to publish something to establish scientificity. Second, the domain itself was professionalizing, producing its own journals. One had to be visible within the nascent community, as Conrad Schlumberger reluctantly conceded in a letter to his American manager: "My brother Marcel and I received an invitation from the Secretary of the American Institute of Mining and Metallurgical Engineers asking me to apply for membership. Motive: publication by C. and M. Schlumberger in their periodical. The cost is $15 annually, plus $20 for initial membership. We would get their publications. I don't personally like these tapeworms very much, they are small but they spread—nor do I like these papers that pile up but no-one reads. Nevertheless if you think that it would be *useful* to join, please send the attached application on. If not, toss it in the waste-paper basket."[20] A third pressure to publish was that this might help establish priority and probity in future court battles. Thus Conrad went to Washington to offer to do some relatively uneconomic work for the Geological Survey with the recompense that "Our work for the Geological Survey would be published, but on the other hand it would give us better standing in the United States. And I think there is going to be a lot of competition and we are going to have to go to court to defend our patents."[21] Cumulatively, then, there was sufficient pressure to force the reluctant company to publish something.

But what should they publish? The immediate tension here was between the desire for secrecy and the need to get into print. We have seen the force of the latter. Here is one indication of the former: "Geophysical methods are only useful insofar as they are secret, since these methods are not patentable."[22] The logic of this statement is that what you cannot patent, you do not publish.[23] Thus at the trial, there was a debate about whether or not Schlumberger's patent used direct or alternating current. One of the witnesses said he had no idea, and elaborated: "Schlumberger has been pretty careful not to give out any detailed information of what they do The articles that have been published in general arc practically reprints of the patent with additional examples of some practical log."[24] The patent itself was (in this case fortunately) ambiguous, and the written record deliberately so.[25] One tactic here was simply to be particularly careful about what one was writing. The American manager E. G. Léonardon summarized the internal discussion about publication just as Schlumberger was becoming successful: "I replied to you that some propaganda with a scientific air is possible, without our necessarily giving away all our secrets. There is no need for us to publish anything of high scientific integrity—they are already calling Mason's talk 'scholarly' in this country." He was not in favor of producing a pamphlet:

The "pamphlet" is, in effect, a piece of propaganda and what is in it is necessarily considered to be advertising. We need to keep people in suspense, to show them that we are out there and that we can write sensibly about physics and mathematics. On occasion, so as not to give our competitors information, we'll have to be jesuitical and lie by omission. I think that we can say a lot without necessarily talking about resistivities and other specialities that we worked hard to get together. All we need to do is publish a few papers that don't have the degree of probity you are so concerned about. Naturally it would be difficult for you to sign them, but no-one is asking you to make that sacrifice.[26]

Rather simpler than tucking away what one wanted to keep secret and printing only what one could defend in a court of law was to request that others do the writing. Thus when De Golyer came to edit a textbook about prospecting, he agreed with Léonardon that rather than call on Schlumberger itself to write about its method, he would ask their customers to give their appreciation of what it was and how it worked.[27] These were the friendly accounts. Naturally, Schlumberger's competitors were also producing official texts— Sean Kelly, for example. Léonardon disapproved of sharing information with Kelly, who used Jakosky's method:

> There is also the way he treats us when he writes a technical article. Shall I mention the fact that in everything he has published during the last three years on the SP phenomenon, he has consistently emphasized the fact that this was an old discovery, dating from the early part of the last century. Also, in an article published on dam sites, he made references to everyone concerned with this important aspect of geophysical exploration except Schlumberger. At least, the slight indications given in this connection seem to imply that we were late-comers in this kind of investigation.[28]

Thus defense of current and future patents was central to Schlumberger's publication strategy. This overriding concern actually extended into internal correspondence too: the decision was made to conduct all correspondence in English (although most of the staff were French-speaking) so that in the event of future use of the records in a court case, they would have documentary evidence in the court's own language. Clearly this comes back reflexively on the historian, since the only material he or she often has at hand is archives that have been written with the official history in mind.

Schlumberger needed to publish to gain respectability and to establish their own version of the history of prospecting; if they said too much, they might jeopardize their patents either by giving up secrets or by specifying the method to such an extent that someone could invent around it. Accordingly, scientific articles and textbook references would be as far as possible inscribed within the account that the patents gave of their technical objects or would be written by others with knowledge of only the public face of the company. Either way, they would necessarily accept the framework that the patents themselves imposed: that the technical objects they protected were indeed black boxes at rupture with the local, and that they constituted a marked progress on past methods.

We can now begin to see how the initial analogy with defending a historical position applies. In each case, there emerges a generally accepted account about "what actually happened." In each case, production of this account is in much the same way severed from its own institutional roots—the preservation of the integrity of scientific/technical knowledge or of academic debate. That is, there is no reference in the accepted account to the variety of ways in which the account got imposed—each party claims to be constructing a discourse of pure truth (though perhaps accusing the others of being impure). There is general agreement about where the high ground is, and we will see how this is constructed in the next section.

Patents and Company Strategy

We have seen that priority was a key issue in determining the validity of a patent. Priority itself was subtly negotiated behind the scenes. One wanted some of it, but not too much. As explained by Léonardon in a memo in 1936, the company could have traced their invention back to 1921, "but in a discussion concerning prior art it would be detrimental to our interests to go back as far as 1921 and take the risk of reducing accordingly the life of our resistivity patent 1,819,923". On the other hand, Schlumberger could use the six-year hiatus between a *precursor* the idea (the surface analogue) to prove that inventive genius was required.[29] However, this later date needed some protection. Therefore Schlumberger acquired the Schlichter patent 1,826,961. It was of limited scope and was never used for petrol, but it did antedate Schlumberger's by a few weeks and involve measurements of conductivity down a drill hole. The company had been advised that the fact of Schlumberger's prior invention was sufficient: "As stated above, it is easy to prove the Schlumberger invention as far back as 1927. However, we did not know if Schlichter could still anteriorize this date. On the contrary, having pooled our interests, we gained the good will of the Schlichter interests and had communication of all the Schlichter early experiments." After the acquisition they also acquired Schlichter's records and learned that April 1928 was the date of the first test.[30] Thus when there was no public trial of priority in the form of a patent battle, there was often a rewriting of history behind the scene: the Whiggish account was as carefully constructed as the object it defended.[31]

Another way in which Schlumberger had to defend its patents behind the scenes was from parasitic attacks. Faussemagne's described one such attack: "The only competition we had at that time was from J. J. Jakosky. He was a professor at the University of

California and his tactic was to examine the patents and see how he could get around them, then to develop some apparatus as cheaply as he could and then sell the patent for the process that the original company had forgotten to take out a patent for."[32] The solution was the same as for Schlichter, but this time constructing the boundary around the patent rather than its priority:

I met Jakosky in 1936 in a field in Louisiana. At that time after we had done our electrical logging, he ran a log with a truck that he had prepared. Jakosky's principle was to measure resistivity with an alternating current and monoelectrodes. His graphs were really disastrous—they were inclined and had no baseline; but he touched them up so that they looked like Schlumberger logs. He was a bit of a pest, and we bought out his equipment.[33]

This parasitic form is a special case of the process of "inventing around" others' patents, which has been studied by Hughes and others.[34]

Let us assume that a given company has patents with a safe boundary in time and space. What do they do with them, and how does this reflect on the printed account of the technical object they are using? Schlumberger was involved in its early years in two major patent battles, with much the same result in each instance. The first was with the Lane Wells company, which did mechanical coring of oil wells. This battle was in one sense similar to that with Halliburton: both Halliburton and Lane Wells were encroaching on Schlumberger territory, one from the direction of oil-well casing and the other from the direction of perforation. The central office in Paris wrote to Léonardon about the threat from Lane Wells:

It should not be forgotten that it is only electrical logging that has rendered the use of the "single column" and of perforation possible. Therefore it would be illogical, and at the same time extremely upsetting, if our organization could not make anything out of this new activity whose success we are largely responsible for. Finally, if we let our competitors occupy the perforation market, they will get themselves a lot of trucks with electrical equipments and other surface installations, and will be naturally tempted to trespass on our own domain and to do electrical logging themselves.

With respect to Lane Wells, headquarters recommended that Schlumberger study their patents on perforation and quickly get an "explosive perforator," using the same principles as Delamare-Mozi, Moza, and Lane Wells, but with as many differences as possible on points of detail from Lane Wells.[35] Just as Schlumberger tried to work around Lane Wells's patent, Lane Wells tried to work around theirs:

"Wells thought that he was going to eat us, even if we made a very expensive arrangement with him because he had his hands on a Swedish patent that was very complicated using alternating currents. With his lack of education in electrical matters, he thought that it was better because it was more complicated. This is more or less the same as what happened before with the Germans, who had wanted to do electrical logging. It often happens. More complex things give more information, but mixed up—and you have to know how to sort it out."[36]

Thus Schlumberger decided that rather than go into a patent battle, which they might lose because of the possible anteriority of the Swedish patent, they could fight in the oil field: "In the meantime, we do not agree to enter into a direct battle in the United States on the patents issue. . . . The fight with Lane Wells in the States should be carried out with the Fuse Perforator, that is to say 'technically.'"[37]

Looking back, one of Schlumberger's early heads saw this as the right decision:

The Lane Wells trial was very positive for Schlumberger, in the sense that it freed us from the sword of Damocles hanging over us in the form of perforation. Schlumberger was still doubtful about starting up perforation in the US without the license for the Lane Wells patent, but Schlumberger's main goal in starting the trial was to prevent or reduce the danger of Lane Wells taking a large part of the logging market. Put simply, Schlumberger's attitude was: it is better to make things more difficult for ourselves with respect to perforation by being forced to pay a license provided we can make it more difficult for Lane Wells to get into the logging market.[38]

The choice was a happy one in terms of future patent battles, since (as noted at the time): "Finally, we shake off the image that we are trying to set up a monopoly, which American courts abhor, and we get official recognition of our patents."[39]

There were, then, two main ways in which the companies involved strove to maintain the possibility of a Whig, internalist account of their inventions. Firstly, they shored up the historical account and the boundaries around the technical objects by buying up patents. This cleared away cases of "real" priority and parasitic attacks. Secondly, when the real world became too messy for such an account to possibly stand up, they settled out of court and fought "technically." In neither case was the purity of the patent's history a self-sufficient objective. As for the academic journals of our analogy with an historical debate, if you wanted to carry on the battle in court, you had to be able to produce such an account. And just as for our analogy, what happened in court was not necessarily the determining factor of success: the account had to be *respectable*, but it made little difference if it was actually *right*. Thus for the case that did go to court, against Halliburton, Schlumberger ultimately lost the case but gained custom and consolidated their de facto at the same time as they lost their de jure monopoly. Henri Doll indicated this in an interview in 1973, when he pointed out that Standard Oil had given Halliburton their logging patent.

The best proof that Standard set up the whole thing to attack our monopoly and not so much to get cheaper logs was that as soon as the Schlumberger patents were decreed to be valueless, Standard gave its business back to Schlumberger so that it would be done better, knowing that Schlumberger could not charge too high a price or sit on their monopoly and say: it's not worth bothering ourselves about doing research.[40]

He concluded, "If I had been in their position I'd have done exactly the same thing—that is to say attacking the monopoly and setting up a competition which could remain latent."[41]

Thus when we look at company strategy with respect to patents, we find that there is no belief on the part of the actors in the independence either of the patent (its history and boundaries are actively constructed) or of the patent trial (which is seen in the Halliburton instance as an integral part of a strategy aimed elsewhere). It is no accident that an untenable Whig, internalist account appears in textbooks and scientific articles: a deliberate filtration process goes on behind the scenes, the result of which is that such an account constitutes the public face of the company.

Again there is a clear analogy with the process of historical debate. For an argument to be accepted by the courts or pass the peer review process, it needs to take an accepted form—to go through a filtering process. This means in particular that reference to any mediation between the "facts of the matter" and the company's/author's personal position is systematically excluded (James Clifford gives a particularly lucid account of this kind of exclusion for shoring up the authority of the participant observe).[42] It is not that Tawney and Trevor-Roper would not recognize the personal and political roots of their controversy; it is just that they knew very well that they would not get published if they made reference to it in their arguments. As a result, their debate was carried out with remarkable vituperation, but with all the animus vested in legitimated forms of disagreement about the interpretation of evidence. For both patents and the historical debate, this exclusion is not a given; it is the end point of the process of deciding where and how to fight the battle. The filtration process is much the same in the two instances: for Schlumberger and for the professional historian, "irrelevant data" includes any account of company/personal motivation or of other battle sites. In both cases a world that no one really believes in (with real truths, arguments being decided purely on their merits) is underwriting a "legal fiction." We will now see what purposes that fiction can serve for the protagonists.

Patents and Oil Fields

The picture that we see emerging from the preceding sections is that although none of the actors believe that technical objects are simple enough to present to the world a clear rupture with other objects and a single moment of discovery, all of them believe that it is worth creating this impression locally and temporarily. In the last section we saw that they made the effort; we will now try to model *why* they do so. We will be looking at the third level of the analogy with the historical debate: the way in which the historical process itself can validate or render irrelevant the historical view that is taken.

Everything revolved around the issue of timing: the complex of relations around the state of the industry, the development of petroleum geology and geophysics and the kind of oil reserves being discovered. The overall process was admirably described in 1937 by J. H. M. A. Thomeer: "Electrical coring and modern oil field exploitation have deeply influenced each other, have grown up together and are now so intimately linked up that separation would be impossible without serious damage to both."[43] In a sense, this was the real motivation for defending patents (and thus blocking competitors for a time). Thomeer is referring here to the fact that Schlumberger influenced the drilling method used, the drillers' mud that circulated in the well, the ways in which fields were exploited, and the use of other prospecting methods. Schlumberger aimed to *survive* long enough in the field so that they could build up the necessary expertise and methods to win, and so that they could transform the field so that they had to win. They needed to "geophysicize" geologists sufficiently: "we had a lot of education and penetration work to do, particularly with geologists"[44]; at the same time as geologists strove to "geologize" the industry: "This conquest of the industry by geology has been not unlike the process of metasomatism, to borrow an appropriate if somewhat imposing term from the hard-rock geologists. Metasomatism is that important process in ore deposition whereby the invading solution, although it leaves the outward form or body of the host rock unchanged, nevertheless entirely transforms its intrinsic character, replacing the original internal constituents, molecule by molecule, with substitutes of its own selection. So has geology reformed the business of producing oil."[45] Comparison can be made here with the steel industry, whose "chemicization" Thomas Misa (this volume) brings out, again in conjunction with the implantation of a new technology.

Looking back on the process, Henri Doll recognized that neither of their two revolutionary methods really did what they said they would: "But then again what does absolute scale mean? The log that you take essentially measures the invaded zone [the zone that has been invaded by the driller's mud]. Now maybe that means something, but after all if you really want to know whether there is water or oil with your resistivity measurements we now know that there is not so much difference between a petrol layer and a water one, unless you want to look behind the invaded zone—which neither of our measurements did."[46] But these methods that did not really work were effective in combination with local experience: "It is clear that ample experience and thorough knowledge of local conditions in a given field are essential factors for a reliable interpretation of Schlumberger logs, the same factors, therefore, as are required for any other method of studying the productive prospects of reservoir rocks. Correct interpretation of the meaning of oil smell, chloroform cuts, sand appearance etc observed on core samples, also is largely a matter of local experience."[47] Or, as Gish (a Schlumberger rival) wrote in an article on electrical methods: "In the present state of the art, success depends to a considerable extent upon familiarity with the method and apparatus, and this is not gained in a day."[48]

This process of building up prior knowledge and expertise (dealing with the messy and the local) went hand in hand with the process of building up the patent (creating purity), as the following samples from internal correspondence show: "When we know the region better, our information will be more sure and precise than that of the geologists, and we will save on many feet of useless forage;"[49] "Petrol, petrol, that is all that interests the companies. Ah, if we knew a procedure for detecting it, they wouldn't find us too expensive. They are always asking me: 'Can you distinguish a petrol-bearing layer from an aqueous one?' The answer is delicate: yes we can, providing you first tell us if there is any sand there at all;"[50] or again the report that in the San Joaquin Valley, "certainly more than in Los Angeles, the operators have to depend on our interpretations since our logs are not so evident by themselves. The small operator in the San Joaquin Valley cannot very well take the risk of a Geoanalyzer survey, which would be valueless to him since the Geoanalyzer engineers can not have the large experience we have already gathered. As for the majors, they are with a few exceptions still using entirely Schlumberger."[51]

What is interesting here is the indication of a double process. In the first process, Schlumberger was defending their patent by claiming that it gave the correct historical account of the development of electrical logging. In the second process, they were changing the nature of well digging so that electrical logging was the only possible adjunct to the drilling process. Thus they were in a messy way creating the hegemony that they already claimed was the correct account—and their claim that this was the correct account helped them create the hegemony. The analogy with the Tawney/Trevor-Roper case is twofold. First, imagine the reception of Tawney's work being such that a number of his readers change allegiance and fight in the class war. His "legal fiction" would have stood up well enough at the time to motivate others to make it true. Second, imagine Tawney participating in an insurrection in England. In so doing he would have been struggling to impose his account of history, and if his struggle succeeded, then his history would prove to have been correct—institutionally and therefore in the history books. In his case, Marxism would be at both the beginning and end of the process. These two counterfactuals illustrate the general point that historical truths change with time, and that historical accounts feed recursively into this change. In Schlumberger's case, an internalist history is at the beginning and the end—at the beginning in terms of the patent and its defense, and at the end because the world of oil will have been sufficiently altered that Schlumberger is inevitable.

Conclusion

Schlumberger engineer Martin recalled in an interview that, "in 1933, Conrad Schlumberger said: 'Basically, what makes this business of electrical logging a difficult one is that we want to go down into other people's holes.' This was certainly true at the time, but a few years later I would say that the companies were desperate to offer us their holes."[52] The sexual metaphor expresses a historical truth: Schlumberger did become inevitable. We have seen that they did so in part by losing a patent battle, in which they tried to defend two patents that did not really work. To explain this paradoxical success, we have seen that there were things going on in at least three different places (the courtroom, the company, the oil field) and at three different rhythms, and that the *timing* of these three processes was at least as important as the "correctness" of the science or the history within any one of them.

We can now return to the two accounts of similar relationships that I referred to in the introduction: Callon's and Shapin and Schaffer's. Callon's actors are unconsciously backing two radically different sociologies; Shapin and Schaffer's are consciously developing their political philosophy in the laboratory. In our case there was no conflict—conscious or otherwise—between the types of historical account that the actors sought to impose. Schlumberger worked away at both ends for a Whiggish, internalist account of their own inventions; the achievement of such an account would be a mark of their victory and would constitute an entry in the history books. We are dealing with the process by which accounts that scientists and technologists give of their own disciplines/crafts are always willing to be externalist about the other and internalist about themselves. It is not just that the internalist explanation is what they "believe" or what will prove to have been true if they succeed. If it works, it also means that they have changed the world sufficiently that it becomes true. Changing the world in this instance means changing the nature of geophysics, the practice of drilling an oil well, and so on—each case is different.

It should be noted, however, that these changes are not limited to changing institutions. In our case, it meant changing that part of the world that oil companies dealt with, and so changing the world of which they were phenomenologically aware. Thus, for example, in the early years there was a popular superstition against faults: "The absurdity, as now we understand it to be, was commonly expressed in the once-familiar words: 'The country appears too much broken up.'"[53] Faulted country was avoided, and where there was an oil field, faults were not part of the model.

We had to work as quickly as possible to show that, contrary to what was the received opinion of all Gulf Coast geologists that there was generalized lenticularity and no faults, that both existed. When I say that the general theory was that there was no possibility of correlation in the Gulf Coast, that was an absolute theory that all geologists in the region held and when I went to visit Monsieur Thomson of Pure Oil shortly after my arrival, I saw a magnificent model of a field with all the productive layers being lenticular. They worked from the wells they were producing from and got lost somewhere in nature.[54]

Faults were good news for Schlumberger: "A fault seems to exist running East-West, the south side being the higher and the one where up to now large production has been found. Besides the major fault, minor unconformities and some lenticularity prevail in the area, and the transition from oil to grey sand takes place sometimes on very short distances. This condition will make the use of electrical logging in this new field entirely systematic."[55] Thus that bit of the world that is faulted was ignored and unmanageable before Schlumberger and was sought after and manageable after.

In general, then, the geophysicists sought to change the world and to change the industry so that their emergent method could develop and become true. It is possible to read the story in realist terms: Schlumberger had discovered a method that on development fit better with the facts than any other. This is structurally not so different from a social determinist position, which would argue that Schlumberger had discovered a (social) method that on develop-ment fit better with the society of the oil industry than any other.[56] It is also possible to read it in constructionist terms: Schlumberger built up the truth of their method using all the tools at their dis-posal—from resistivity meters to rhetoric to rationalism. My own feeling is that the constructionist reading does not go far enough in recognizing that both physical and social reality exerted a definite influence: Schlumberger's work was significantly shaped by both, even as it was shaping. The realist/social determinist account clearly goes too far the other way.[57] What we are left with is a situation in which "nature" and "society" are both emergent realities that are constructed by their components at the same time as their compo-nents construct them.

There is a two-part strategy to making things more real within this emergent reality. First, an appropriate space is created by manipul-ating awareness contexts. Official debate and institutional battle are kept sufficiently unaware of each other that a series of publications and secondary institutions can build up in blissful ignorance of the messy side to such an extent that the two levels become, to borrow a somewhat weary phrase, semi-autonomous.[58] There is an inside and an outside—the former occupied by company engineers, strategists, and scientists, and the latter by the public face of science. Bruno Latour's mask of Janus, with one face before the fact is created and another after it is accepted, comes to mind.[59] Here we have shown that in industrial science, this "before" and "after" occupy separate spaces; they become "inside" and "outside." Once the externalist explanation has been sufficiently constrained that only the patent holders believe it (and then only in private), then the boundary between inside and outside is in place.[60]

There is, then, no need to maintain this space forever. The second part of the strategy involves maintaining it only *long enough* that social and physical reality will alter.[61] This is what we characterized as the issue of timing. If Schlumberger survived for long enough in the patent battle, then whether or not they won that battle, they would win out historically, because they would have created the breathing space within which to impose their reality on the oil field. And that is exactly what they did.

Notes

This essay was made possible by a generous grant from the Fondation les Treilles. I would like to thank Leigh Star for her incisive comments.

1. Noble 1977, 111.

2. Hughes 1983.

3. Reich 1985.

4. Dennis 1987; Bowker 1989.

5. Cambrosio, Keating, and Mackenzie, forthcoming; cf. also Pinch, Ashmore, and Mulkay (this volume).

6. I am using classical definitions of "Whig" and "internalist." Whig history refers to an account of uninterrupted progress (as opposed to a discontinous/revolutionary or a nonprogressivist account); internalist history refers to a belief that scientific/ technological change can be explained in its own terms (without an "externalist" reference to its social context).

7. For a development of this concept, see Strauss 1978 and Glaser and Strauss 1966.

8. For Trevor-Roper, see Trevor-Roper 1951 and 1953. For Tawney, see Tawney 1960. For an account of the debate, see Russell 1973 and Stone 1972. Stone asserted, "There have been few more brutally savage attacks in academic journals than that in which H. R. Trevor-Roper demonstrated the exaggerations and inconsistencies in my first article."

9. See, for example, Stone 1973.

10. Smith (1974 : 260) gives a good general account of the process by which the historical account is formed in rupture with historical reality:

We begin at what actually happened and return via the social organization to 'what actually happened'. ... The two social organizations, of the production of the account and of the reading of the account, are distinguished in the first place because at the point at which the account is put into its final form it enters what I shall call 'document time'. This is that crucial point at which much if not every trace of what has gone into the making of that account is obliterated and what remains is only the text which aims at being read as 'what actually happened.'

11. Callon 1986b.

12. Schaffer and Shapin 1985.

13. Bowker 1987 and 1988. For a good technical history, see Allaud and Martin 1976.

14. Bowker 1987.

15. Compare Misa (this volume), who points out that participants in his dispute were ever ready to identify the "interests" of the others.

16. Cf. Lyell's *Principles of Geology*, for example, wherein all manner of social and ideological motives are imputed to everyone before Lyell himself.

17. Heiland 1940, 623.

18. Gish 1947 (1932), 498–499.

19. Box Pechelbronn, Folder 1922. [The early years of the Schlumberger archives are partially housed at the Ecole des Mines, Paris. They are divided into folders within subject boxes, and I annotate them accordingly.]

20. Box USA 1926–1933, Folder 1930, 5/8/30, Conrad Schlumberger to E. G. Léonardon.

21. Box USA 1926–1933, Folder 1926, 3/11/26, Conrad Schlumberger to Paris.

22. Box Pechelbronn, Folder 1925, Letter to Conrad Schlumberger, 6/6/25.

23. Compare the problem of secrecy as tied to classified publications. The story of the computer expert who was not allowed to read his own paper has reached the proportions of academic myth. In general, even though Hughes and others have highlighted the problem of secrecy in industrial science, it has not received systematic attention—in large part because the interesting documents are secret.

24. *Schlumberger* v *Halliburton*, vol. 13, 1205.

25. For the value of ambiguous patents, see the story of Diesel as recounted by Thomas (1987). Diesel had to defend a theory of his engine that he knew to be incorrect because it was enshrined in an overly precise first patent.

26. Box USA 1926–1931, Folder 1927, 23/11/27, Léonardon to Paris. Cf. the remark by Léonardon that he liked a conference memo by Charrin, but that it gave away too much and should not be distributed too widely. He proposed a copy of his English-language pamphlet: "whose commercial and non-scientific aspect will displease you." (Box USA 1926–1931, Folder 1927, Léonardon 27/6/27).

27. Box USA 1934–1938, Folder 1934, E.G. Léonardon, 3/8/34.

28. Box USA, Folder 1934–1938, E.G. Léonardon to Paris Office, 23/1 2/36.

29. Box Historique, Folder Operations, Léonardon, 8/9/36.

30. Box Historique, Folder Operations, Léonardon, 8/9/36.

31. Misa (this volume) also points to the variability of the category of "real priority."

32. Box Interviews 1, Interview Faussemagne. [The interviews I cite were carried out in the 1970s, largely by Martin in the preparation of his and Allaud's book or by A. Gruner-Schlumberger, daughter of Conrad Schlumberger (one founder of the firm).]

33. Box Interviews 1, Interview Faussemagne, p. 15.

34. See Hughes 1983 and Reich 1985 for example.

35. Box USA 1934–1938, Folder 1934, 14/6/35 Paris to E. G. Léonardon.

36. Box Procès, Folder Interview with H. G. Doll.

37. Box USA 1934–1938, Folder 1936, 29/5/36, SPE to Léonardon.

38. Box Procès, Folder Interview with H. G. Doll.

39. Box Procès, Folder Echange de Lettres à propos de Halliburton.

40. Box Procès, Folder Interview H. G. Doll.

41. Box Procès, Folder Interview H. G. Doll.

42. James Clifford, "On Ethnographic Authority," Representations, 1, 2, 1983.

43. Typescript by J. H. M. A. Thomeer of Bataafsche, "The Application of Schlumberger Electrical Logs in Oilfield Operations," in Box USA 1934–1937, Folder 1937.

44. Box Interviews 3, Interview Mathieu.

45. Pratt 1940, 1211

46. Box Procès, Folder Interview with H. G. Doll.

47. Typescript by J. H. M. A. Thomeer of Bataafsche, "The Application of Schlumberger Electrical Logs in Oilfield Operations," j in Box USA 1934–1937, Folder 1937, p. 6.

48. Gish, 1947 (1932), 502.

49. Box USA 1926–33, folder 1929, 11/12/29, Gallois to Henquet.

50. Box USA 1926–33, folder 1929, 17/11/29, note by Deschâtre.

51. Box Notes Techniques 1930–1938, "California Progress Report for the months of October, November and December 1936," p. 25.

52. Box Interviews 3, Interview Martin.

53. Clapp, 1929, 686.

54. Box Interviews 3, Interview Mathieu.

55. Box Progress Reports 1936–1944, "California Progress Report for the months of October, November, and December, 1936," p. 10.

56. This comes out clearly in our analogy with the Tawney/Trevor-Roper debate; since one way in which their positions are developed in the real world is by changing the structure of society and thence the craft of historian—so in this case "realism" is conflated with "social determinism."

57. Latour 1987.

58. To put it another way, we can refer again to Smith's (1974) analysis, wherein two different social worlds are created: that of the *author* (Schlumberger here) for whom the whole process is open and messy and that of the *reader* (the courtroom, the science student) for whom the process is closed and clean.

59. Latour 1987.

60. Pinch, Mulkay, and Ashmore (this volume) elucidate a similar coexistence of different technological rhetorics—a "strong program" for one audience and a "weak program" for another. In their case, the former rhetoric appeals to economists and government ideologues, the latter to administrators. Thus even when these rhetorics

cohabit the same text, the one group will come away with one inside story and the other with an opposed one. That ambiguity is deliberately fostered in order to be able to black box the technology—and once it is black-boxed, the strong program will doubtless become the dominant rhetoric.

61. Compare here Law and Callon's fruitful notion of negotiation space, which operates similarly in a slightly different context (Law and Callon this volume and Callon and Law 1989).

3

The Social Construction of Fluorescent Lighting, Or How an Artifact Was Invented in Its Diffusion Stage

Wiebe E. Bijker

Technology is assumed to be designed, developed, and produced by engineers.[1] They are at the drawing boards and behind the laboratory benches; they apply for patents, model the prototype, and test in the pilot plant; they show the newly born artifact to the press and, if lucky, they figure prominently in the glossy photographs of stories about heroic inventors. Once these engineers have produced the technology, it is passed on to the sales people, the managers, the trade, and, finally, to the users. Engineers design technology, managers produce it, salespeople sell it, tradespeople distribute it, users use it. Alas, this neat and orderly image of technical development, so pervasive in all but the most recent technology studies, is not only too simple—it is wrong.

This chapter has two aims. First, I want to show that the application of a linear stage model of technical development is detrimental to understanding the development of technical artifacts. Rather, no stages can be distinguished. I will demonstrate how the modern fluorescent lamp was designed during what commonly would have been called its "diffusion stage." If the fluorescent lamp is considered a static artifact, forever fixed and unchanging since it left the General Electric laboratories on April 21, 1938, it is difficult to understand what actually happened and the original lamp's relation to the present fluorescent lamp. Instead, I will analyze, from a social-constructivist perspective, the fluorescent lamp as something that was continually reshaped and redesigned by the various social groups involved.[2] The second aim of this chapter is to provide an illustration of the possibilities of integrating the social-shaping and the social-impact perspectives on technology.[3]

Part of the development of the fluorescent lamp is described in detail, using the social constructivist approach (SCOT).[4] In the SCOT descriptive model, *relevant social groups* are the key starting point. Technical artifacts do not exist without the social interactions within and among social groups. The design details of artifacts are described by focusing on the *problems and solutions* that those relevant social groups have with respect to the artifact. Thus, increasing and decreasing degrees of stabilization of the artifact can be traced. A crucial concept in SCOT (as well as in the Empirical Program of Relativism, EPOR, in the sociology of scientific knowledge, to which SCOT is closely related) is *interpretative flexibility*. The interpretative flexibility of an artifact can be demonstrated by showing how, for different social groups, the artifact presents itself as essentially different artifacts. The theoretical concept of *technological frame of a social group* is employed to explain the interactions within and between social groups that shape the artifacts; these technological frames shape and are shaped by these interactions (Bijker 1987).

Relevant Social Groups

It is relatively easy to identify the relevant social groups by "following the actors".[5] They are themselves quite explicit about it. For example, Howard W. Sharp, utility executive and member of the Lamp Committee of the Edison Electric Institute, used the phrase "I have delayed replying ... in order to coordinate with the rest of the boys,"[6] referring to what I will call the social group of utilities. Other social groups are clearly identified as well: "It is apparent that dealing with the fixture manufacturers, as a group, involves delicate negotiations."[7] Historical actors sometimes even seemed to be anticipating the problems of historians and sociologists of technology and, in the case of the fluorescent lamp, deliberately tried to maintain their group's integrity. For example, Sharp wrote, in connection with the fluorescent lighting developments:

It is quite desirable that we maintain the united front that has been established so far in connection with this light source [... and] concerted action on the part of responsible people in the lighting business is necessary in order to prevent "runaways."[8]

Actors' accounts may correct the researcher's intuitions. For example, in 1984 I employed the fluorescent lamp as an "obvious" example of a technical development where it would not be useful to consider a separate social group of women: neither for the actors, nor for me as analyst, would that provide any further insight—or so I thought then. (Pinch and Bijker 1984, 415). However, O. P. Cleaver, a leading Westinghouse executive, thought otherwise when he analyzed the problems in the home lighting field with respect to fluorescent lighting:

The widespread acceptance of fluorescent lighting in the home will depend directly upon the housewife, who is generally alert to new ideas that give comfort to her family and beautify her home, provided the cost does not exceed the family budget—and more important, provided she is made conscious of the advantages of the new equipment through national advertising and neighborly example.[9]

This executive clearly recognized the social group of women as relevant for the development of the fluorescent lamp.

Actors provide an effective starting point from which to identify relevant social groups. In that sense, "relevant social group" is an actor category. However, it is indeed only a starting point, and this method is not proposed as an "idiot-proof" recipe for carrying out a social constructivist case study. Several methodological issues are still unsolved. For example, it may be difficult to decide whom to treat as spokespersons for a specific relevant social group, although this will, again, often become clear if we let the actors speak for themselves. In some instances—for example, when one social group is splitting into two—groups may not accept someone acting as its spokesperson, but that will again become evident by "listening" to the actors.[10] Another problem is that only "vocal" attributions of meaning are analyzed, and there is always the danger of the analyst not "hearing" the voices of some parties. This ethnographic approach deliberately focuses on meanings attributed to artifacts and does not take the route of imputing hidden interests to social groups as, for example, Marxist structuralism or Parsonian functionalism would do.

After following the actors, the second step in identifying relevant social groups is what might be called "historical snowballing."[11] While following the actors by reading historical documents, the researcher notes each actor and every social group that is mentioned. Subsequently those new actors and social groups are also followed, and at some point no more new names or social groups will be encountered. Of course this is an ideal sketch, because the researcher will have intuitive ideas about what set of relevant social groups is adequate for the analysis of a specific artifact and, consequently, will not follow this road to its very end. This methodological model serves here primarily to argue that there is no essential problem involved in using the concept "relevant social group" in empirical (whether sociological or historical) studies of technology.

The problem of delineating relevant social groups (and, for example, deciding whether it is more effective to use two different groups rather than one) is still a matter for the intuition of the researcher. Obviously, the list of relevant social groups that results from this strategy needs to be simplified and ordered. To start with, many actors may be taken together to form one relevant social group, but then some of the groups thus created may turn out to be too large. For example, in the case of the bicycle it was decided that a separate group of women cyclists needed to be incorporated in the description (Pinch and Bijker 1984). Similarly, in the later stages of the fluorescent lamp case, the social group "government" had to be split into the Antitrust Division of the Department of Justice and the War Department. Also here, an important starting point is to let the actors speak for themselves.

"Relevant social group" is both an actor and an analyst category. When following the actors in their identifications, definitions, and delineations, it is the actors' relevant social groups that we get. The central claim in the social construction of technology is that these relevant social groups *are also relevant for the analyst*—"relevant social group" is also an analyst concept.

I will now describe the relevant social groups in the fluorescent lamp case. Only two social groups will play an important role in this chapter—the Mazda companies and the utilities. The other groups will be described briefly.

The Mazda Companies

The social group of Mazda companies consists of General Electric and Westinghouse.[12] They were, at that time, commonly referred to as "Mazda companies" after their incandescent lamp trademark "Mazda." Through its licensing system, General Electric had control of about 90 percent of the incandescent lighting market during the period 1913–1940 (Rogers 1980). The General Electric patent-licensing system consisted of two classes of licenses. The class A license was granted only to Westinghouse. It gave the licensee the right to produce a certain percentage of General Electric's own lamp output and, among other things, the right to use General Electric's Mazda trademark. Licensees with a class-B license were allowed to produce a smaller quantity of certain types of lamps, and they could not use the trademark. Hygrade Sylvania Corporation was such a class-B licensee.

A crucial role in maintaining this almost absolute control of the lighting business was played by the intimate connections between the lamp manufacturers (General Electric and Westinghouse) and the electricity-producing utilities. This is a specific example of the general observation that relevant social groups do not only constitute themselves, but they also help to maintain other social groups and the relations between them. The basis of the relations in this case was an understanding that each side would work in the interests of the other. The utilities undertook to sell and promote Mazda lamps— and the appliances and other electrical apparatus of the Mazda manufacturers as well—and, for their part, the Mazda manufacturers undertook to promote their products in such a way as to add to electricity consumption. The Mazda companies also supported and participated in campaigns and programs conducted by the utilities to increase the use of electricity supplied by them. For example, in the 1930s a large number of utilities gave their customers free renewals of lamps of higher wattage to keep their sockets filled. The lamps used in these campaigns were Mazda. General Electric and Westinghouse supplied the lamps at reduced prices to the utilities, with free renewals. The intimacy of the relations between the Mazda companies and the utilities is evident in that the utilities not only sold Mazda lamps, they also advertised them and promoted their use.

The Utilities

Obviously, the social group of utilities is going to play a prominent role in this story. Who were they? Each utility was a private company, operating one or several central stations to generate and sell electricity. The utilities had a number of strong collective organizations and can be seen as acting, through these organizations, as one social group. The utilities, although ordinarily independent of each other, did act in concert in matters affecting their common interest. For instance, over one hundred utilities belonged to the Edison Electric Institute (E.E.I.). Another large organization of utilities was the Association of Edison Illuminating Companies (A.E.I.C.). Each of these associations extended to every part of the country. The E.E.I. and A.E.I.C. were made up of committees and groups, composed of representatives of the member utilities, who among other things handled policies for the industry. The policies were either determined at the meetings of the organizations as a whole or formulated by the particular committees themselves, on the basis of their knowledge of the desires of the industry. Frequently this knowledge was derived from questionnaires sent out to all utilities.

Two important committees were the Lighting Sales Committee (E.E.I.) and the Lamp Committee (A.E.I.C.). These committees had over many years worked very closely with representatives of General Electric and Westinghouse in setting policies with respect to the manufacture, distribution, and use of (incandescent) lamps manufactured by the two Mazda companies, and the promotion of such lamps by the utilities. The Electrical Testing Laboratories also played an important role. This organization was owned by the utility companies and engaged in commerical testing of electric lamps and other electrical equipment.

The Fixture Manufacturers
The social group of fixture manufacturers deserves separate mention. In the fields of both incandescent and fluorescent lighting, the Mazda companies produced mainly lamps. Sockets, reflectors, and other kinds of auxiliaries were produced by smaller companies. For incandescent lighting, a system of specifications had been set up, and fixture manufacturers had to design their products according to those specifications. Their products were tested by the Electrical Testing Laboratories. A similar plan was to be developed in the field of fluorescent lighting.

The Independents
The social group of independents consisted of lamp manufacturers not bound to General Electric by patent licenses in the fluorescent field. Hygrade Sylvania, a B licensee of General Electric in the incandescent lamp field, was the only company in this social group. According to its B license, Hygrade Sylvania was allowed to produce 9.124 percent of General Electric's net sales quota in incandescent lamps (Bright and Maclaurin 1943). With only about a 5.5 percent market share, Sylvania did not have a great stake in the incandescent field. It acquired a patent position on fluorescent lamps to counter that of General Electric and Westinghouse. Hygrade Sylvania started production of fluorescent lamps in 1939 and soon had 20 percent of the fluorescent market. The company had developed its fluorescent lamp independent of General Electric and was, in this field, not bound by license agreements—hence the name "independent."

The Customers

The social group of customers does not have its own direct voice in this story. However, the results of market research conducted by the utilities and the lamp maufacturers reveal some of the attitudes of this social group. Also, an analysis of the popular technical press may reveal parts of the meanings as attributed by the social group of customers, since this press may be considered to reflect the views of customers.

The Government

A peculiar role in the fluorescent case is played by the social group of the government—or more precisely, by two groups: the Antitrust Division of the Department of Justice, which filed an antitrust suit against General Electric and Westinghouse in 1942; and the War Department, which asked the Attorney General to make an application to the court for an adjournment because such a trial would seriously interfere with General Electric's contribution to the war effort.[13]

The Interpretative Flexibility of the Fluorescent Lamp

The interpretative flexibility of the fluorescent lamp can be demonstrated by showing how different relevant social groups attributed different meanings to it, constituting three quite different artifacts in the period 1938–1942: the "fluorescent *tint* lighting lamp," the "high-*efficiency* daylight fluorescent lamp," and the "high-*intensity* daylight fluorescent lamp." The first two artifacts played an important role in the "load controversy" between the Mazda companies and the utilities. The third artifact was at the same time instrumental in and resulting from reaching closure in this controversy.[14]

On April 21, 1938, the fluorescent lamp was released commercially by the Mazda companies, General Electric, and Westinghouse. These "fluorescent lumiline lamps" were explicitly aimed at "tint lighting." The new lighting device could provide brighter and deeper colors of a wider variety than was previously possible with incandescent lamps. Because of their ability to produce "light in hitherto unobtainable pastel tints as well as pure colors,"[15] they were expected materially to affect many phases of lighting practice. Moreover, although their installation costs were higher, they were thirty to forty times more efficient than incandescent lamps for color lighting.[16] Lighting applications mentioned ranged from theater interiors to ballrooms, from specialty shops to art galleries, from showcases to game machines, from railway cars to homes. Some of the applications suggest that the Mazda company executives were already thinking of general indoor lighting, but this is not very explicit.

In these early days of fluorescent lighting, the lamp was a "fluorescent *tint* lighting lamp" for the relevant social group of utilities, just as it was for the Mazda companies. This is not surprising, because the utilities' knowledge of these lamps was rather limited and based almost exclusively on information provided by the Mazda companies. The new lighting device was introduced in a way that did not suggest any revolutionary change in lighting practice. Three utility men remembered the occasion:

Its presentation was as casual as developments in incandescent sources were wont to be. There was the usual amount of discussion, but the impression seemed to be that here was a light source rich in color and high in efficiency, but low in total light output, expensive, and generally suitable for only special applications.[17]

Thus even when daylight lamps were discussed, this was done in the context of special purposes and tint lighting, as is clear from a memorandum of the Chairman of the A.E.I.C. Lamp Subcommittee:

The daylight tubes it is to be anticipated will have most utility. Because of the small wattages and small production of heat these lend themselves particularly well to showcase illumination. Because of the white light they should find large application for color matching purposes.[18]

The origins of this specific artifact, the "fluorescent *tint* lighting lamp," can be traced back to the 1939 New York World's Fair. Of course, the standard histories of discharge lighting in general, and of fluorescent lighting more specifically, go back to the Geissler tube (1860), the Moore tube (1895), the Cooper-Hewitt lamp (1901), the Claude neon tube (1912), and the Risler, Küch, and Holst lamps (1920s to 1930s).[19] Often, these histories are presented in the perspective of a quest for general indoor white lighting. Considering what we know now about the presently stabilized usage of fluorescent lamps (i.e., general indoor daylight lighting), it is intriguing why that first artifact was the fluorescent tint lighting lamp and not immediately the other lamp that eventually stabilized: the high-intensity daylight fluorescent lamp. The fluorescent *tint* lighting lamp seems to be a strange deviation from the (retrospectively apparent) linear path, which ran from the goal of general white indoor lighting to, at its end, the artifact high-intensity daylight fluorescent lamp. The actors show how to understand this detour by guiding us to the World's Fair.

Ward Harrison, engineering director of the incandescent lamp department of General Electric and most prominent spokesperson of the Mazda companies in the early days of fluorescent lighting, admitted,

There were a couple of World's Fairs in the offing that were going to be lighted almost entirely with the high tension tube lighting if they were not supplied with some lamps of ours.[20]

Other relevant social groups also saw the World's Fairs as the reason for dragging the fluorescent lamp "out of the research laboratories by a caesarian operation."[21] As the fixture manufacturers described this episode retrospectively,

The pressure of the demand for a new illuminant to be exploited at two World's Fairs was too much [for conservative judgment to prevail]. The 15- and 20-watt fluorescent lamps were produced for use at the Fairs— others wanted them—and a new illuminant, with a lot of unexplored implications, was launched.[22]

This view is confirmed by the lighting engineers of the World's Fair themselves (Engelken 1940). This context makes the emphasis on tint lighting understandable. In the $150 million transformation of 1,200 acres of salt marsh and wasteland into the New York World's Fair, so vividly described in the novel by Doctorow (1985), color schemes of architecture and artificial illumination played an important role:

A zoning and color scheme adopted prior to the construction insured architectural unity, and harmony of plan, design, and treatment throughout the whole area.... The color scheme ... is coordinated with the physical layout. Starting with white at the Theme Center, color treatments of red, blue or gold radiate outward with progressively more saturated hues. Adjoining hues blend circumferentially along the avenues. The illumination was fitted to this scheme [so] as to maintain the basic pattern by night as well as by day, but with new and added interest and charm after sunset. (Engelken 1940, 179)

Obviously, tint lighting was an important objective for the lighting engineers who were designing the first large-scale applications for these fluorescent lamps.

But within half a year of the introduction of the fluorescent tint lighting lamp, another artifact emerged: the high-efficiency daylight fluorescent lamp. A flood of advertising over the signatures of the major lamp companies streamed out, containing such statements as, "three to two hundred times as much light for the same wattage," "cold foot-candles," "amazing efficiency," "most economical," and "indoor daylight at last." The utilities started to fear that the high efficiency of the fluorescent lamp might reduce their electricity sales. As the utility executive Carl Bremicker of the Northern States Power Company said about his utility employees, "They had better get their white wing suits ready because very shortly General Electric and Westinghouse would have them out cleaning streets instead of selling lighting."[23] An internal Westinghouse memorandum lends support to the utilities' worries. It concluded that "the average utility lighting man sees in the rise of fluorescence a decrease in his relative importance."[24] The memorandum presents a comparison of the profits, based on a 4 cent rate and with equal costs to the user. The design data were unfavorable to fluorescence—almost any other selection would have emphasized the differences. The result of the comparison was that, to the utility, fluorescence was only half as important as incandescence; to the lamp suppliers it was six times as important, to the equipment manufacturers three times as important, to the contractor 20 percent more important.[25]

Thus a controversy developed—the "load issue." It took the form of a competition between the two fluorescent lamp artifacts. The utilities, having been alerted by their discovery of the high-efficiency daylight fluorescent lamp, tried to keep the other artifact, the fluorescent tint lighting lamp, in the forefront. They argued that claims about high efficiency were true, but only when fully qualified. And this, they claimed, was not done. For example, when the "three to two hundred times as much light" statement was accompanied by the picture of an office, the customer might expect amazing efficiencies. And this, the utilities argued, could have been true only if that customer was willing to have green or blue light.[26] The utility lighting staff was irritated by this misleading publicity, and in trying to fill in the rest of the story found that they were being immediately accused of excessive self-interest. They resented their position of apparently throwing cold water on fluorescent lighting because they were trying to tell the complete story. Long and detailed arguments were given to point out that the high-efficiency daylight fluorescent lamp really did not exist, but that it was mistaken for the fluorescent tint lighting lamp, which indeed was a valuable new lighting tool, but only for limited purposes.[27]

The principal spokespersons for the Mazda companies did not agree with the conclusion that the load on the electricity networks would fall, thus decreasing the utilities' profits. And so they continued to push, albeit carefully, the high-efficiency daylight fluorescent lamp. Harrison, for example, was convinced that only in some instances would consumers cut down on electricity use, but that, on average, their electricity consumption would go up.[28] However, the Mazda companies had their own problems with the high-efficiency daylight fluorescent lamp: at the moment of its commercial release, there was no known relation between life and efficiency in fluorescent lamps; in fact, the life of the lamp was not known. They knew that it was something more than 1,500 hours when the lamps were given their original rating, but they did not know whether it could work out to be 15,000 hours or much more. As Harrison said to an audience of utility executives, "Instead of having 93 per cent of our business in renewals in good times and bad, it may be that our first sale will be almost our last sale to a given customer."[29] Nevertheless, the Mazda companies were developing a more differentiated line of fluorescent lamps because, as Harrison explained in 1940,

The effect of changes in the efficiency of fluorescent lamps, changes in their rated life and changes in price have radically affected their over-all operating costs, so that in twelve months ... [these changes have] brought the lamp more seriously into the field of general lighting.[30]

Obviously, the artifact he was describing was the high-efficiency daylight fluorescent lamp, not the fluorescent tint lighting lamp.

The controversy was fierce, probably because the relevant social groups of Mazda companies and utilities both felt that their common control of the lighting market, as exerted in the incandescent era, was at risk. This threat became especially acute when a third relevant social group entered the arena—the independents, notably the Hygrade Sylvania Corporation. In late 1939, the Mazda people started to worry about Hygrade Sylvania:

There are figures which seem to indicate that the Hygrade Company is selling as many fluorescent lamps as General Electric and Westinghouse combined. Apparently, they are going out and "beating the bushes," so to speak, installing sockets in the smaller companies on main streets throughout the United States.[31]

The aggressive sales policy employed by Hygrade Sylvania created as much of a problem for the utilities as it did for the Mazda companies. The utilities sensed a realignment of forces taking place among the lamp manufacturers. Hygrade claimed to have basic patents for the manufacture of fluorescent lamps and did not recognize the patents held by the Mazda companies. The utilities feared that this realignment of forces, together with the competitive situation that attended it, might lead to methods and activities that would disorganize the whole lighting market "to the detriment of the public and the utilities who were standing on the sidelines." That Hygrade Sylvania was capturing a sizable portion of the market was claimed by the company and acknowledged by the Mazda people.[32] Hygrade Sylvania clearly was advancing the high-efficiency daylight fluorescent lamp, although downplaying the economic risk for the utilities. For example, the Hygrade manager W. P. Lowell, before an audience of utility and Mazda company executives, argued in answer to the question why fluorescent lighting was demanded by the public:

Why is it demanded? For many reasons: its daylight color, soft quality, reduced shadows, novelty (it's new, modern, smart), real or imaginary economy. But don't worry too much about those who think they are saving money by using fluorescent lighting to save a few watts. If the overall value—combining the sheer dollars and cents with all other qualities—if the net value is not right, the product will fall of its own weight. You can't fool all the people all the time.[33]

Thus, Hygrade Sylvania's activities resulted in pouring oil on the fire.

Various ways of closing this load controversy were tried. One was a certification plan for fluorescent lamp fixtures. With such a certification scheme the Mazda Companies hoped to stimulate and control the production of fixtures by the auxiliary manufacturers and thereby to check the growth of Hygrade Sylvania, which was producing its own fixtures. The realization of this certification plan took a long time because the specifications initially proposed by the Mazda companies were unacceptable to the utilities; only after negotiating for almost a year, could the specifications be agreed upon. Then it only further consolidated the closure of the load controversy, which had by that time been reached through another process. This other process was the design of a new fluorescent lamp—the high-intensity daylight fluorescent lamp. In the next section I will describe this closure process.

Stabilization of the High-Intensity Daylight Fluorescent Lamp: Changes in Technological Frames

To understand how closure was reached in the controversy between the Mazda companies and the utilities through the design of the high-intensity daylight fluorescent lamp, I will describe the changes in the technological frames of both groups. These technological frames will be sketched by focusing on three of their dimensions: goals, current theories, and problem-solving strategies. The fluorescent technological frames of the Mazda companies and the utilities were quite similar but for two or three crucial differences relating in particular to the goals and problem-solving strategies.

The utilities' main goal was to sell electricity, whereas the Mazda companies' goal, in the context of this study, was to sell lamps. Left at that, this would be a rather trivial observation. However, goals do not straightforwardly define the actions taken by the relevant social groups. The respective technological frames influence, for example, the way these goals are translated into problem-solving strategies.

The theoretical base of the Mazda companies' fluorescent frame was formed by electricity and gas discharge physics, whereas the utilities obviously used, primarily, power electricity physics. Neither played an explicit role in the historical episode I describe here. The utilities' frame was supplemented by what they called the "science of seeing," which focused on the quality of lighting, including such things as brightness, contrast, shadows, diffusion, and various kinds of glare. This theoretical part of the utilities' frame did play a role: emphasis was placed on seeing and the prescription of lighting that would contribute maximum visibility to the task. As the utility people said themselves, rather pretentiously, about the years of incandescent lighting: "A true Science of Seeing was born. . . . It was here that the Cooperative Better Light–Better Sight Movement was started, and lighting practice became firmly entrenched in the philosophy of 'results to customers.'"[34]

The last words in this quotation hint at an important element in the problem-solving strategy of the utilities: they pictured themselves as servants of the public, or even teachers of that public. Thus an important goal was to increase public confidence in lighting technology and to promote (the utilities' version of) knowledge about that technology. In this context, the utilities highly valued cooperation with the Mazda companies:

The lighting industry, based upon a sound Science of Seeing and united by the Better Light–Better Sight Movement, has presented a solid front to the public. This has captured the interest of strong professional groups, increased the customer confidence so important to future growth, and has proved successful commercially.[35]

The implication for the technological frame of the utilities is that, when confronted by a problem, their standard strategy was to reformulate the problem as educational—and hence to design better advertising strategies and sales methods. This is what happened in the case of the load problem. Talking about the public, which was thinking about lighting costs in terms of current costs instead of "true costs," they formulated as their task "to educate them properly to the true cost and value of adequate lighting [, which] is not an easy job."[36] It is important to see that this problem-solving strategy was not the only one possible. Another strategy would have been, for example, to define appropriate standards and impose them on other relevant social groups, thereby solving the problem. The utilities did indeed adopt this strategy, but only as a second choice at a relatively late stage, when the Mazda companies had already proposed the certification scheme for the fixture manufacturers.

After this brief characterization of the two technological frames, we will resume the story where we left off: early in 1939, when the load controversy took the shape of a conflict between two competing artifacts—the fluorescent tint lighting lamp and the high-efficiency daylight fluorescent lamp. During the first year after the commerical release of the fluorescent lamp, the tension increased between the Mazda companies and the utilities.[37] A dissociation of the cooperation established in the incandescent lighting era seemed not unlikely. Mueller, Sharp, and Skinner remembered: "The question was quickly asked . . . : could it be that the sound principles of the Science of Seeing so assiduously promoted were built upon sand, to be cast aside at the first gust of commercial expediency?"[38]

To settle this conflict, a conference of representatives of the utilities and the Mazda companies was held on April 24 and 25, 1939, at the headquarters of the General Electric Lamp Department at Nela Park, Cleveland. The utility representatives referred to this meeting as "the fluorescent council of war."[39] At this conference the idea emerged that fluorescent lighting might be reserved exclusively for high-level lighting. Retrospectively, one can argue that a third fluorescent lamp was designed—not on the drawing board or at the laboratory bench but at the conference table. This artifact—the high-intensity daylight fluorescent lamp—came slowly into being during this meeting, as is apparent from the minutes:

> There was considerable discussion on the outstanding features of fluorescent light with particular reference to daylight quality. Some thought that low footcandles of daylight fluorescent lighting made a person appear sallow— on the other hand, 100 or more footcandles in the Institute Round Table Room (previously inspected) seemed satisfactory to everyone. From the discussion, it was generally agreed that 50 to 100 footcandles of fluorescent lighting could readily be installed without creating any impression of high level lighting. At least in some instances it was believed that 50 footcandles of fluorescent lighting would appear like no more illumination than 25 footcandles of filament lighting.[40]

Now, what could be expected to happen to this idea? Considering the utilities' technological frame, it is understandable that the situation was perceived in terms of advertising. It was decided that the use of fluorescent lamps for general lighting would not be emphasized "until commendable equipment is available giving 50 to 100 footcandles levels." This decision clearly demonstrates the effect of the specific problem-solving strategy in this technological frame. Instead of treating the problem primarily as one to be solved by advertising and educating, it would have been conceivable to treat it as, for example, a mainly technical problem—concentrating all efforts on the development of lamps and fixtures to provide high-intensity lighting. Indeed, quite the contrary happened, as I will try to show.

In line with their technological frame, the utilities pressed the Mazda companies to adopt specific ways of advertising the fluorescent lamps, and they were quite satisfied with the result. After a difficult start on the first day, the second day's discussions produced what utility executives saw as "a most complete capitulation."[41] Mueller thought he understood how closure was reached:

> I think it was probably due to the fact that they realized they were definitely on unsound ground the way they had been operating, and they also knew ... that the utilities realized it and were going to do something about it, and they knew that they really couldn't put across any lighting promotion without the help of the utilities. They were anxious to settle these matters with our group, because they thought that we were in the best position to get something in return for their capitulation.[42]

The large lamp companies issued statements of policy concerning the promotion of fluorescent lamps and tried to implement the new policy in all parts of their organizations. For example, in the "statement of policy" by General Electric, issued officially on May 1, 1939, the company conceded that

because the efficiency of fluorescent lamps is high, it might be assumed that the cost of lighting with them is less than with filament lamps; as often as not this conclusion is erroneous. The cost of lighting is made up of several items—cost of electricity consumed, cost of lamp renewals, and interest and depreciation on the investment in fixtures and their installation. All of these factors must be properly weighted to find the true cost of lighting in any given case. The fluorescent Mazda lamp should not be presented as a light source which will reduce lighting costs.[43]

Similarly, the Westinghouse statement read in part, "We will oppose the use of fluorescent lamps to reduce wattages."

Mueller believed that one of the most important results of the conference was that the lamp companies seemed inclined to take the utilities into their confidence, as part of the lighting industry, in the development of promotional plans, instead of "shooting the works" first and then letting them know about it.[44] The Mazda companies clearly had the same ideas as the utilities about the need to reach an agreement. According to J. E. Kewley, manager of the lamp department of General Electric, "The ... statement of policy [was] issued particularly to allay the fears of the utility companies." And E. H. Robinson, another General Electric official, viewed the policy statement as a declaration by the lamp department signaling "Here's how we stand, boys, we'll play good ball with you central stations but we'll expect the same brand of ball from you too" (Committee on Patents 1942, 4772). Thus, the agreement on the new high-intensity daylight fluorescent lamp not only solved the load controversy but also saved the cooperation between the two important relevant social groups.

One would imagine that this must have been quite a successful lamp to have had such an impact on the two most powerful social groups in the electric lighting business in the United States. This was not the case, however, at least not in any straightforward way. The lamp did not even exist. According to Walker, the antitrust division attorney, there even was no immediate prospect of fluorescent lamps (or any other kind) that would give anything like 50 footcandle levels. The average with incandescent lamps in 1939 was probably about 15 footcandles, and no single installation gave anything like 50. Nevertheless, the impact of this artifact, the social construction of which started at the Nela Park conference, was not small. Ironically, part of its impact at that conference may have been caused by its not yet being available, as Walker argued:

The reason why the utilites did not want the fluorescent lamps promoted until they . . . would give 50 to 100 footcandles levels of lighting was that the utilities felt that if they could ever get fluorescent lamps of intensities that strong, fluorescent lamps would then use so much electricity that the utilities would not suffer as a result of the fluorescent lamps replacing the incandescent lamps. (Committee on Patents 1942, 4771)

The new General Electric and Westinghouse policy statements were not widely broadcast, and it is not difficult to guess why the public was not informed about the cancellation of the high-efficiency lamp and the effort to sell the high-intensity lamp instead (Committee on Patents 1942, 4773).

Thus, the utilities' technological frame (partly) shaped the fluorescent lamp. On the other hand, as result, the technological frames of the utilities and the Mazda companies had to change to adapt to this new artifact, the high-intensity daylight fluorescent lamp.[45] And so the fluorescent lamp had a social impact in turn. For example, an adaptation of the theories element in the utilities' technological frame was one of the first effects. After the agreement at the Nela Park conference, the utilities immediately started to elaborate on the idea of high-intensity lighting. Two days after the conference, a note was written by the Electrical Testing Laboratories for the A.E.I.C. Lamp Subcommittee arguing for daylight lighting by providing a theoretical evolutionary/biological justification:

It will be noted that our eyes have evolved under the brilliant intensity of natural light in the daytime and under the dull flow of firelight in the evening. There is some reason to think that with light of daylight quality people will not be satisfied with the low intensity of illumination which is more or less acceptable in the case of light of warmer tone as that of tungsten filament lamps. Where the daylight lamps are to be used, the logical procedure is to work toward the equivalent of daylight illumination, which at once moves practice into higher ranges of illumination intensity.[46]

In a later report, this argument was pursued further. It was claimed that lighting research indicated that the human eye functions more naturally above 100 footcandles than under 15 to 50 footcandles— considered the upper limit of most incandescent general lighting systems at the time. The ultimate advantage of fluorescent lighting to the consumer was, therefore, to be found in properly designed installations giving at least 100 footcandles. Experience with the user's reaction to general lighting from sources of "natural" daylight quality indicated, it was said, a preference for daylight quality if high intensities were provided. In an E.E.I. memorandum, an elaborate argument was forwarded to explain why this leap to 100 footcandles was not as big as it seemed—and, indeed, was quite necessary for fluorescent lighting:

Lighting of substantially daylight quality, when appraised by the eye, appears to be much less than equivalent footcandles of light from normal incandescent sources. The reasons for this are scientifically and psychologically obscure, but the fact remains that general satisfaction with lighting is based in large measure upon the user's appraisal of the amount available, and as such must be taken into account when applying light to large areas. Furthermore, the light from the "colder" tube appears blue and depressing at low intensities and produces an uncomplimentary effect upon goods or people in commercial or work areas. This effect disappears at levels of illumination above 100 foot-candles.[47]

Thus the utilities' technological frame was adapted to the new high-intensity daylight fluorescent lamp.

The stabilization of this lamp did not come about smoothly. Neither party to the Nela Park agreement adhered to it without occasional lapses, and in particular, the utilities felt that the Mazda companies were regularly violating the agreement in their advertising. On May 24, 1939, Sharp wrote to Harrison that utility employees had complained to him about a display in the General Electric building at the New York World's Fair. This display purportedly consisted of a 20-watt fluorescent lamp and a 20-watt incandescent lamp, with a footcandle meter showing how much more light was given by the fluorescent than by the incandescent lamp. Objecting to General Electric having this display on exhibit in their building at the World's Fair, Sharp stated,

If this demonstration is as explained to me I think it does violate the sprit of the understanding that our group had in Cleveland. As a matter of fact, I would think it violated the fundamental concept of the lamp department that advances in the lighting art should not be at the expense of wattage but should give the customer more for the same money. I hope you can find a way to change this exhibit, so that it does not give misleading impressions to the crowd who will see it.[48]

Harrison replied to Sharp that the exhibit was not intended to demonstrate the amount of electricity that could be saved by the use of fluorescent lamps, and that the exhibit was being withdrawn.[49]

I have discussed one adaptation of the utilities' technological frame—the addition of a theoretical explanation of the need for high-level lighting. Another adaptation of the technological frames of both utilities and Mazda companies further enhanced the stabilization of the high-intensity daylight fluorescent lamp and thereby contributed ending the controversy between the Mazda companies and the utilities. This adaptation involved the development of a standard method for comparing the costs of incandescent and fluorescent lighting. These "standard cost comparisons" are analogous to testing procedures used in engineering. Such testing procedures, if they exist, form an important element of technological frames.[50] In this case, it was not easy to reach agreement on such a standard method; in part, the cause of the problem was that this generation of lighting people had little experience with competitive illuminants. The incandescent lamp had been well-nigh universal, so that lighting design principally involved technical considerations, with relatively simple arithmetical calculations of equipment cost. Now that there was a light source as radically different as the fluorescent lamp, lighting design involved a more complicated cost comparison before it became clear which source would best meet specific requirements.[51]

However, an even more serious barrier to an agreement on standard cost comparisons were intrinsic differences in interests between the two parties. First, there was a difference in focusing on the costs of electricity versus focusing on the costs of the apparatus. For the Mazda companies, it was attractive to emphasize the low cost of electricity and disregard the high price of the apparatus itself, whereas for the utilities the opposite was true. Secondly, the utilities' primary aim in developing a standard method of comparing lighting costs was to pursue the fight against the high-efficiency lamp. The Mazda companies, despite their "capitulation" at the Nela Park conference, were of course not anxious to help the utilities in that fight.

Late in 1939 the E.E.I. Lighting Sales Committee did propose a standard method, which it claimed to be universal in application and to ensure an evaluation of all factors. Utility lighting people seem to have been almost unanimous in their approval of this method, whereas manufacturers gave only lukewarm assent. Utility executives commented on this lack of enthusiasm by the Mazda companies:

Their reluctance is founded on the fact that true cost calculations bring out the items of high fixed charges and expensive fluorescent lamp renewals.

These are customarily slighted by manufacturers' representatives and job-bers in their eagerness to bring out unquestioned reductions in energy cost, foot-candle for foot-candle. Wide experience with the use of this method in investigating fields of fluorescent application have shown that no blanket statement as to cost can safely be made. As often as not, when a true cost comparison is made on a five- or six-year depreciation basis, the fluorescent installation is more expensive for the customer than filament incandescent lighting. This clearly points out that it is fallacious to sell fluorescent light-ing on the basis that it is the most economical form of lighting.[52]

But, apparently, there was not much choice open to the Mazda companies: some months later Mueller could come to the conclusion that "this method possibly cannot be dignified by being called an 'industry standard', [but] it comes pretty close to that. It has also been endorsed by the lamp companies and is used by them."[53] Thus, the development of this cost comparison method as a new element in the utilities' technological frame strengthened their struggle against the high-efficiency lamp and contributed to the stabilization of the high-intensity lamp.

And indeed, after their initial hesitations, the Mazda companies decided that the utilities' promotion of the high-intensity lamp could be profitable to them as well. Harrison, arguing to a utility audience, observed that if they would just substitute the fluorescent lamp for incandescent on a candlepower-for-candlepower basis, in the long run they might wind up with less lamp business: "Only by using fluorescent lamps to at least double the present standards of illumi-nation can we hope to get renewal business enough to make it worthwhile for us—and then the lamp will be valuable to you."[54] And General Electric developed a new line of fluorescent lamps of higher wattages, thus giving physical existence to the high-intensity lamp at last.

Dynamics of Technological Development: Interactions between and within Relevant Social Groups

The social construction of the high-intensity daylight fluorescent lamp took place in a situation in which two technological frames were dominant. Elsewhere I have argued that in such a situation *symmetrical amortization* or *amalgamation of vested interests* is one of the possible stabilization processes.[55] Indeed, if we take the phrase "am-ortization of vested interests" in its true heterogeneous sense (as compared to its common, narrower, financial definition), it provides an adequate characterization of what happened in the fluorescent lighting case. The Mazda companies dropped the high-efficiency lamp and agreed to restrict themselves to making the high-intensity lamp. On the other hand, the construction of the high-intensity lamp certainly was not a complete victory for the utilities. Mueller clearly viewed the Nela Park agreement as a compromise when he argued the need for the E.E.I. Lighting Sales Committee to make some additional concessions to the Mazda companies:

Unless our committee does something now to give them [i.e., the Mazda companies] some publicity on their change of pace, and to get the utility industry as a whole interested in the promotion of fluorescent lighting along sound lines, I think they will drop us and either try to get action through some other body, or else come out with another "To Hell With The Utilities" campaign, and go it alone, knowing that they have quite a strong customer appeal in their efficiency and novelty story.[56]

And so the utilities started slowly to adapt their policy toward advertising the fluorescent lamp, switching "from informing to selling" in their fluorescent lighting presentations. Thus the conflict was indeed solved by a piecemeal adaptation by both parties to the new situation: amalgamation of vested interests.

Until now I have treated the utilities and the Mazda companies as monolithic entities. However, the pressures from outside caused by the process of closing the load controversy created tensions within both organizations. For example, within General Electric there was opposition to the Nela Park agreement. The lamp department, which had participated in the Nela Park conference, experienced resistance within the large General Electric organization. When the General Electric Supply Corporation published a catalog listing and picturing fixtures unequipped with shielding, Harrison (of the lamp department) objected because "the repercussions from central stations are likely to be formidable."[57] The catalog showed fixtures both bare and equipped with shields. However, all the listed prices applied to the bare lamp fixtures only; the shield was shown as an extra item, requiring separate and additional catalog numbers when ordered. Then the statement appeared that the use of shields would result in 30 percent less light. It is evident that this way of presenting the fluorescent lamps would stimulate customers to buy the un-shielded lamps, thus getting more light out of the lamp for the same amount of electricity. Harrison threatened: "Of course, it is up to the General Electric Supply Corporation ... to formulate their own policies, but I do not think that a penny of Lamp Department money should be spent to support a campaign of this kind."[58]

In its answer to Harrison's letter, the G.E. Supply Corporation justified its advertising on the grounds that it was necessitated by Hygrade Sylvania competition.[59] The background of Harrison's threat was that the Nela Park conference had been that fluorescent lamps would not be installed without "proper shielding." In the case of incandescent lighting, shielding was necessary to avoid glare. This was hardly the case with fluorescent lighting, but evidently shielding would decrease the net light output.[60]

Tension like that within General Electric is likely to arise between actors with different degrees of inclusion in one technological frame (Bijker 1987). The G.E. Supply Corporation was bound to have a relatively low inclusion as compared to the G.E. lamp department because the latter was more intimately involved in the establishment of the new fluorescent frame of the Mazda companies, in which the selling of only the high-intensity lamp was the goal, and which was aimed at nursing the collaboration with the utilities. For the sales-people of the Supply Corporation, the "old" incandescent techno-logical frame of simply selling as many lamps as possible, and thereby competing with other lamp manufacturers, was more prominent.

Similarly, such tensions can be observed within the group of utilities. For example, Bremicker (of the Northern States Power Co.) wrote to Mueller, after having received a report on the Nela Park conference, that this was not enough: he wanted a specific retraction from the Mazda companies stating "that fluorescent lighting is not known to be applicable for any lighting purposes except colored or atmospheric lighting and certain phases of localized lighting such as wall cases, showcases, display niches."[61] Bremicker concluded that he did not want the utility companies to be hoodwinked into a cooperative program of promoting fluorescent lighting. The position of Bremicker was similar to that of the General Electric Supply Company, in that he did not attend the Nela Park meeting and, hence, was only marginal in the newly established technological frame. Although Bremicker was the only member of the Lighting Sales Committee of the E.E.I. who did not endorse the results of the conference, this may not lead us to the conclusion that his critique was entirely exceptional. Sharp proposed to Mueller (both were participants in the Nela Park conference) not to send out the entire minutes of that meeting. Instead, a letter with only a brief outline should be sent out, which "would indicate that the committee is still on the job, [and which would] serve to keep the utility group united, and give our committee some additional backing from the field, thereby making it harder for anybody to divide our forces.[62]

Sharp recognized the potential tension between the highly included participants of the Nela Park conference and the other utilities executives with a much lower inclusion.

Conclusion

The first point I wanted to illustrate with this case study is the saliency of the social constructivist approach in understanding the development of artifacts in their "diffusion stage," as it is called in the "old" technology studies. To understand the design process of technical artifacts, we should not restrict ourselves to the social groups of design-room engineers or laboratory personnel. Basic to all "new" technology studies is the observation that even in the diffusion stage, the process of invention continues.

In demonstrating the interpretative flexibility of the fluorescent lamp, it became clear that, after the official release of the lamp and thus during its diffusion stage, there were at least two different artifacts. In this first step of the SCOT model I showed that "laws of nature," or the claim that "it is working," did not unequivocally dictate the form of this artifact. Thus it was clear that something more was needed to explain the constituency of the fluorescent lamp. In the load controversy that originated from the competition between these two artifacts—the tint-lighting fluorescent lamp and the high-efficiency daylight fluorescent lamp—closure was reached by designing a third artifact, the high-intensity daylight fluorescent lamp, as a kind of compromise. The specific form of this invention in the diffusion stage could be explained while making the second point of this chapter.

This second issue I wanted to address concerns the integration of the social shaping and social impact perspectives on technology. One of the key elements in recent technology studies can be captured by the "seamless web" metaphor: the development of technical artifacts and systems should be treated as if technology and society constitute a seamless web. Indeed, historians and sociologists of technology are trying to reweave the web of technology and society in such a way that they can avoid traditional categories such as "society" and "technology" altogether (Hughes 1986a; Bijker, Hughes, and Pinch 1987b). Thus, for example, the social shaping of a technical artifact and the social impact of that technical artifact are to be analyzed with the same concepts, within the same frame and, preferably, even within the same study.[63]

A concrete example is the high-intensity daylight fluorescent lamp. I have tried to show how this artifact emerged from the social interactions between the Mazda companies and the utilities during their efforts to reach closure in the load controversy—thus the high-intensity lamp was indeed socially shaped. On the other hand, this artifact also influenced society by giving rise to new lighting standards which in the end became universally accepted—so this artifact also had quite a social impact. Offices, for example, turned into potential surgical suites from which, after 1974, one could remove up to half of the original fluorescent tubes without any damage to the clerks' working conditions. How can both sides of the coin, both faces of the Janus head (Latour 1987), both parts of the seamless web, be described and analyzed within one conceptual frame?

To capture this double-sided character of technological development, I have employed the concept "technological frame." The technological frame of a social group is shaped while an artifact, functioning as exemplar, further develops and stabilizes within that social group—the social impact side of the coin. But a technological frame in turn also determines (albeit to different degrees, depending on the degree of inclusion different actors have in that frame) the design process within that social group—the social shaping side of the coin. Thus forms the concept "technological frame" a hinge between the social impact and the social shaping perspectives on technology.

Notes

1. I am grateful to David Edge and Robert Frost for their stimulating comments on different drafts of this chapter. This research was funded by the Netherlands Organization for Scientific Research (NWO), grant 500-284-002.

2. This chapter is mainly based on one source: the hearings held before the Committee on Patents of the U.S. Senate, August 18, 1942 (Committee on Patents, 1942). This committee had a specific political mandate to investigate possible violation of antitrust regulation by General Electric, Westinghouse, and the electric utility companies. Especially the contributions by John W. Walker, attorney of the Antitrust Division, Department of Justice, and the questions asked by the committee's chairman, Senator Homer T. Bone, do reveal some bias in this respect. This, however, does not affect my use of this source, because I used primarily original documents, reproduced as evidence in the hearings. References to these hearings will be made, where appropriate, by giving the number of the exhibits of evidence, most of which were presented to the committee by Walker.

3. For a discussion of this problem, see MacKenzie and Wajcman 1985, Hughes 1986, and Bijker, Hughes, and Pinch 1987b.

4. See Pinch and Bijker 1984; the relevant part is also published in Bijker, Hughes, and Pinch 1987a, 17–50.

5. I borrowed this term from Latour (1987).

6. Howard M. Sharp to John Mueller, letter dated April 6, 1939 (Walker Exhibit No. 137), 4993.

7. Howard M. Sharp to Ward Harrison, letter dated March 26, 1939 (Walker Exhibit No. 143): 5000.

8. Ibid.

9. O. P. Cleaver, Westinghouse Lamp Division, Westinghouse Elec. & Mfg. Co., "Fluorescent Lighting in the Home Field," paper presented at the Industrial Conference on Fluorescent Lighting, March 22, 1940, Chicago, Ill. (Walker Exhibit No. 41): 4903.

10. There are some theoretical and methodological problems connected to this issue of spokespersons that I will not discuss in this paper. Identifying spokespersons is one of the central methodological problems at this moment in studying sociotechnology. See, for example, Latour 1987.

11. Harry Collins (1981a) used the snowballing method in his sociology of scientific knowledge studies to identify the core-set.

12. Using a single name to label a social group seems to suggest that this group is a monolithic entity. As will become clear later in this chapter, this is typically not the case.

13. Henry L. Stimson, Secretary of War, to the Attorney General, letter dated April 20, 1942 (Committee Exhibit No. 21): 5030.

14. For an introduction of the term "closure" in the context of studying scientific and technological controversies, see Pinch ad Bijker 1984.

15. Harrison and Hibben (1938, 1530); see also Inman and Thayer 1938.

16. This efficiency, the 'overall efficiency" specifying the efficiency of a lamp to transform electrical power input into light output, was measured in the units "lumens/watt" or "lightwatts/electric watt" (Moon 1936). However, the unit "footcandles/watt" was often used, which is, strictly speaking, not right: footcandles were the unit for the illumination of a surface, whereas lumens were the unit for luminous flux from a lamp.

17. J. E. Mueller, H. M. Sharp, and M. E. Skinner, "Plain Talk About Fluorescent Lighting," paper presented at the 55th Annual Meeting of the Association of Edison Illuminating Companies, January 15–19, 1940 (Walker Exhibit No. 4): 4803. Mueller was Manager of Commercial Sales, West Penn Power Company and Chairman of the E.E.I. Lighting Sales Committee; Sharp was Manager of the Lighting Bureau of the Buffalo Niagara & Eastern Power Corporation; Skinner was Vice-President of the Buffalo Niagara & Eastern Power Corporation; Sharp and Skinner were members of the E.E.I. Lighting Sales Committee as well.

18. Preston S. Millar, "Advanced Memorandum for Meeting of Lamp Subcommittee, May 27, 1938" (Walker Exhibit No. 8): 4821.

19. See, for example, Bright 1949.

20. W. Harrison, talk, probably to a E.E.I. Lamp Committee meeting in fall 1939 (Walker Exhibit No. 80): 4945. He is referring to the high-voltage discharge lighting (for example, neon tubes), mostly employed for outdoor advertising purposes. The history of the introduction of neon discharge lighting into the United States by the

French company Claude Neon and the subsequent negotiations between Claude and General Electric are an interesting part of the prehistory of fluorescent lighting, but will not be discussed here.

21. A. F. Wakefield, The F. W. Wakefield Brass Co., "The Objectives of the Fleur-O-Lier Association," paper presented at the Industrial Conference on Fluorescent Lighting, March 22, 1940, Chicago, Ill. (Walker Exhibit No. 41): 4900.

22. "The Object of the Fleur-O-Lier Association," synopsis of suggested presentation before E.E.I. Sales Meeting, spring 1940 in Chicago (Walker Exhibit No. 104): 4961–4962.

23. Quoted by D. W. Prideaux, Incandescent Lamp Department General Electric Company, to A.B. Oday, Engineering Department General Electric, letter dated February 1, 1940 (Walker Exhibit No. 111): 4973.

24. Westinghouse Commerical Engineering Department to Westinghouse Lamp Division, letter dated July 12, 1939 (Walker Exhibit No. 6): 4818–4819.

25. Ibid. The comparison results in the following table:

For every dollar the user spends annually with incandescence,
the utility gets	80 percent
the contractor	10 percent
the equipment suppliers	6 percent
the lamp suppliers	4 percent

For fluorescence, the dollar is divided into
the utility	44 percent
the contractor	12 percent
the equipment suppliers	20 percent
the lamp suppliers	24 percent

26. Mueller, Sharp, and Skinner: 4803.

27. See, for example, J. L. McEachin to G. E. Nelson, letter dated Dec. 15, 1939 (Walker Exhibit No. 59): 4921; and H. Restofski, Sales Promotion Manager, West Penn Power Company, Pittsburgh, Pa. to James Kernes, Chicago, Ill., letter dated May 7, 1940 (Walker Exhibit No. 92): 4953–4954.

28. W. Harrison, "The Need for More and Varied Types of Fluorescent Equipment," paper presented at the Industrial Conference on Fluorescent Lighting, March 22, 1940, Chicago, Ill. (Walker Exhibit No. 41): 4896.

29. Harrison talk, fall 1939 (Walker Exhibit No. 80): 4945.

30. W. Harrison, "The Need for More and Varied Types of Fluorescent Equipment."

31. O. P. Cleaver, Westinghouse Electric and Manufacturing Company, Commerical Engineering Department, internal memorandum dated April 25, 1940 (Walker Exhibit No. 95): 4955.

32. J. E. Mueller, H. M. Sharp, M. E. Skinner, "Today's Fluorescent Lighting Situation," prepared for the Sales Executives' Conference, Association of Edison Illuminating Companies, Hot Springs, Virginia, September 30 to October 3, 1940 (Walker Exhibit No. 5): 4816.

33. W. P. Lowell, Jr., Hygrade Sylvania Corp., "Industrial Applications of Fluorescent Lighting," paper presented at the Industrial Conference on Fluorescent Lighting, March 22, 1940, Chicago, Ill. (Walker Exhibit No. 41): 4899.

34. Mueller, Sharp, and Skinner, "Plain Talk About Fluorescent Lighting": 4803.

35. Ibid.: 4802.

36. Ibid.: 4807.

37. This tension resulted not only from the load controversy, described previously, but also from a second controversy: the power factor issue. This latter conflict over low power factor will not be discussed in this paper.

38. Mueller, Sharp, Skinner, "Plain Talk About Fluorescent Lighting": 4803–4804.

39. Sharp (Walker Exhibit No. 21): 4848.

40. Draft of detail minutes of the Nela Park conference, April 24–25, 1939 (Walker Exhibit No. 19): 4846.

41. Jim Amos, quoted by J. E. Mueller in a letter to H. M. Sharp, dated May 29, 1939 (Walker Exhibit No. 32): 4858.

42. J. E. Mueller to H. M. Sharp, letter dated May 29, 1939 (Walker Exhibit No. 32): 4858.

43. "Statement of Policy Pertaining to Fluorescent Mazda Lamps," published by the General Electric Company Incandescent Lamp Department in *Lamp Letter* No. S-E-21A (Superseding S-E-21), May 1, 1939 (Walker Exhibit No. 19): 4841.

44. J. E. Mueller to H. E. Dexter, Commercial Manager Central Hudson Gas & Electric Corp., letter dated May 11, 1939 (Walker Exhibit No. 30): 4855.

45. The mechanism of the emergence of a technological frame "around" an artifact as exemplar has been introduced in the context of Bakelite; see Bijker 1987.

46. Notes from the Electrical Testing Laboratories prepared for the A.E.I.C. Lamp Subcommittee Meeting, May 18, 1939, April 28, 1939 (Walker Exhibit No. 135): 4991.

47. Lighting Sales Committee E.E.I., "Recent Developments in Fluorescent Lighting and Recommendations for the Immediate Future," supplemental to report of April 1939 (Walker Exhibit No. 39): 4872.

48. H. M. Sharp to W. Harrison, letter dated May 24, 1939 (Walker Exhibit No. 49): 4916–4917.

49. W. Harrison to H. M. Sharp, letter dated June 1, 1939 (Walker Exhibit No. 50): 4917.

50. It has been argued that testing procedures are an important focus for technology studies. See Constant 1983, Pinch and Bijker 1984, and MacKenzie 1989.

51. Mueller, Sharp, and Skinner, "Plain Talk About Fluorescent Lighting": 4806–4870.

52. Ibid.

53. J. E. Mueller, West Penn Power Co., "The Economics of Fluorescent Lighting," paper presented at the Industrial Conference on Fluorescent Lighting, March 22, 1940, Chicago, Ill. (Walker Exhibit No. 41): 4884.

54. Harrison talk (Walker Exhibit No. 80): 4945.

55. See the "third type of technological development" discussed in Bijker (1987, 184–185). An example of "symmetrical amortization of vested interests" was given

by Thomas Hughes when he described the closure reached in the controversy between AC and DC systems of electricity supply, the "battle of the systems": "it ended without the dramatic vanquishing of one system by the other, or a revolutionary transition from one paradigm to another. The conflict was resolved by synthesis, by a combination of coupling and merging. The coupling took place on the technical level; the merging, on the institutional level" (Hughes 1983, 120–121). Misa (forthcoming) uses the term "amalgamation of vested interest" because of its less narrow economic connotations.

56. J. E. Mueller in a letter to H. M. Sharp, dated May 29, 1939 (Walker Exhibit No. 32): 4858.

57. W. Harrison to N. H. Boynton and E. E. Potter, letter dated May 20, 1940 (Walker Exhibit No. 52): 4917–4918.

58. Harrison talk (Walker Exhibit No. 80): 4945.

59. W. Booth, Manager Lighting Sales, to W. Harrison, letter dated May 28, 1940 (Walker Exhibit No. 53): 4918.

60. This strategy of emphasizing the installation of "proper shielding" offers an example of the social shaping of technology and of the inadequateness of an explanation of technological development which is based on the assumption that "the best working artifact will be chosen." Here, lamps and fixtures were designed to limit light output without much reason or technical benefit (as in the case of incandescent lighting). And although we may now recognize this as the best solution in terms of the utilities' technological frame, it obviously was not the best in any "objective" sense.

61. C. T. Bremicker to J. E. Mueller, letter dated May 16, 1939 (Walker Exhibit No. 31): 4856.

62. H. M. Sharp to J. E. Mueller, letter dated May 22, 1939 (Walker Exhibit No. 34): 4860.

63. Of course, one could say that an author subscribing to the objectives of a seamless web approach should not be writing a sentence like this, using phrases such as "technical artifact" and "social impact." I think that such pedantic criticism is unfruitful. The substantial methodological challenge is to develop analytical concepts that will allow us to realize the aim of analyzing technology and society in such a "seamless web way," but on our way toward that goal one has to make do with what there is—using common language, but as carefully as possible.

II

Strategies, Resources, and the Shaping of Technology

Technologies have social implications. Indeed we have argued that it is impossible to pry technical and social relations apart. The shaping of a technology is also the shaping of a society, a set of social and economic relations. This means that many—perhaps all —technologies are born in conflict or controversy. Different social groups have different concerns, or simply different practices, and hope for or expect different things from their technologies. How are conflicts resolved? How are new technical and social relations set in place? How is irreversibility achieved? The papers in the first section offer certain suggestions. In particular, they point to the importance of the strategies deployed by heterogeneous engineers—for instance, the ways in which system builders deploy organizational and legal resources as they attempt to stabilize a network of social and technical components. The papers in this section build on this theme.

Misa takes us to the history of steelmaking to describe the way in which two controversies were resolved. The first concerned pneumatic steelmaking and a conflict between two groups, each of which held patents crucial to the process. The result was that neither was able to build an advanced Bessemer converter. To have done so would have infringed the patents of the other group. In the geophysical case described by Bowker, Schlumberger defended its patents as a delaying tactic. Although it knew that these would probably turn out to be indefensible, the object was to maintain its strategic position close to the oil exploration companies long enough to build up a body of expertise and a set of practices in which its products were seen as indispensible. Patents thus took the form of a crucial resource. In the case described by Misa they were equally important, but were used quite differently. Instead of fighting in the courts, the two groups agreed to a legal and organizational innovation—the formation of a patent pool from which *both* would profit. The individual legal and technical resources of the two groups were thus combined.

Misa's second controversy concerns the distinction between "iron" and "steel"—one that was important to different protagonists in different ways. Thus, at least in the early stages of the controversy, "steel" carried a price premium. In addition, scientific and professional reputations were at stake: a distinction based on the percentage of carbon demanded the use of (professionally administered) chemical and physical measurements. Finally, there were issues of daily practice. Thus steelmen tended to talk of "steel" to describe metal that fused completely during the process of production, and saw little reason to change their practice. Unlike the patent pool, this controversy was not settled by legal or organizational innovation.

Rather, as the circumstances changed ("steel" ceased to command an economic premium), the inertia of the steelmen carried the day. As Misa puts it, stabilization owed "less to written authority than to the daily practice of thousands of steelmen."

Scientific and professional knowledge, daily practice, and organizational arrangements—all of these also play a role in the story about the technological handling of radioactive waste described by de la Bruhèze. Indeed, de la Bruhèze's account in many ways reads like an essay in technologically informed bureaucratic politics. The name of the deadly game he describes was the mobilization of bureaucratic and organizational resources in order to define the appropriate social and technical arrangements for the handling and treatment of radioactive waste.

Leaving aside its intrinsic importance, there are several striking features of this story. One has to do with the way in which boundaries are drawn. Thus de la Bruhèze illustrates the way in which so much of the bureaucratic maneuvering turned around questions about who or what should have the right to speak, and what they should be allowed to say. The right to speak was, of course, precisely what was at stake. Whoever could speak for the AEC—and make it stick—would define its policy. Accordingly, there were endless tussles about such matters as committee membership and the circumstances under which different individuals and agencies might make their views known. The processes of boundary negotiation described in de la Bruhèze's paper thus resonate with those found in the studies by Misa (who should have the right to speak about the proper character of steel?), Bowker (who should have the right to speak for geophysics, who should have the right to work alongside the oil companies?), Bijker (who should be allowed to define the proper character of fluorescent lighting?), and Law and Callon (who should be allowed to comment on, and make decisions about, the progress of an aircraft project?).

A second feature of de la Bruhèze's story concerns time. None of the protagonists, even in their own estimation, believed that they had a complete solution to the problem of radioactive waste at hand. At best, they believed that they had found methods that would, if properly developed, lead to such a technical solution. Their object, then, was to deploy and freeze organizational and bureaucratic arrangements that would generate technical solutions—and, at the same time, to use the *promise* of new technologies to fix current social relations.

We have encountered this form of bootstrap sociotechnical engineering in the case of the TSR.2 described by Law and Callon. But the geophysical case described by Bowker is also similar. Here Schlumberger believed that a workable new technology might evolve if the current legal challenges, however well founded they might be, could be held off for long enough. To be sure, there is nothing inherently obnoxious about this kind of circularity. But it does indicate that technological innovation may start neither with invention (technology push) nor with consumer demand (demand pull) but rather in an interactive, time-dependent, process of sociotechnical bootstrapping in which promises about technologies and social relations are played off against one another in the search for durable solutions.

If promises are a crucial resource in sociotechnical maneuvering, then the third paper in this section considers another type of resource—the simplifying cultural and cognitive models or strategies used by innovators. Carlson examines the moviemaking endeavors of Thomas A. Edison and his company—efforts that finally failed with the withdrawal of Edison from the moviemaking business. Carlson's argument is that Edison's style of invention and innovation was production-oriented. Thus he tended to create capital goods for business markets, rather than consumer goods for the general public. This worked well in such cases as the telegraph and business equipment, the electric light, and the phonograph. In the case of moviemaking, though, the business was in the end undermined by the development of a consumer culture. In this the hero was not the hard-working inventor but movie and sports stars. Furthermore— and this lay at the root of his business failure—a mass audience grew up that sought diversion, entertainmen, and glamour in its movies, rather than education, information, and "improvement."

In his paper Carlson draws on the sociology of scientific knowledge, and in particular on the notion of the "frame of meaning" developed by Collins and Pinch. This is close to Bijker's concept of "technological frame" except that it applies to engineers and managers alone, and not to other social groups. Nevertheless, like "technological frames," "frames of meaning" are a tool for making sense of the strategies of entrepreneurs. They include cultural commitments, class biases, business strategies, and methods of design. Accordingly, they bring a concern with "narrow" technical factors together with "broad" macrosocial and cultural considerations. Like patents and organizational arrangements, frames of meaning may thus be seen as a resource —a more or less satisfactory set of cultural and cognitive assumptions for making sense of and operating on the sociotechnical world.

4

Controversy and Closure in Technological Change: Constructing "Steel"

Thomas J. Misa

"A bar of steel" is, in the present state of the art, a vastly less definite expression than "a piece of chalk."

Alexander Holley (1873b, 117)

Controversies in science and technology are not new, but recent studies of their genesis and resolution have generated fresh insights into the making of "objective culture."[1] Empirical research now stands behind the view that nature does not determine the form of scientific facts or technological artifacts and that their shape is negotiated among actors. Indeed, the principle of "interpretive flexibility" forms the core of the social constructivist research program. Its advocates maintain there is nothing in principle that cannot be disputed, negotiated, or reinterpreted—in short, become the subject of a controversy. Yet if everything were endlessly negotiated, the effort might exhaust the time and resources of actors and render change impossible. To effect change, actors deploy strategies to hold in place otherwise wayward elements. Through these efforts actors ensure that controversies—if sometimes lengthy—are rarely interminable. In fact, a distinctive characteristic of scientists and technologists is their ability to resolve controversies and engineer consensus. This ability vests facts and artifacts with authority and permanence. What could be tractable or "soft," the topic of interminable controversy, becomes obdurate or "hard," a part of constructed reality. The concept of closure helps account for this remarkable shift.

Recent studies of closure have extended its meaning beyond the familiar one of the effective termination of a controversy.[2] Closure has come to mean the process by which facts or artifacts in a provisional state characterized by controversy are molded into a stable state characterized by consensus. At least four research programs use this concept—if not with identical vocabulary. Working from the empirical program of relativism, Collins (1981a; 1985, 90–100, 142–152) has focused on the principal actors in scientific controversies and posited the "core set" as the group of scientists who achieve closure through argument and negotiation. For Collins, core sets are able to lend methodological propriety to social contingency and thus to certify new knowledge. Advocates of the social construction of technology translate the core set as "relevant social group," controversy as a series of destabilizing problems, and closure as the "stabilization" of artifacts. For social constructivists like Pinch, Bijker, and Elzen, stabilization occurs when and if a social group and an artifact meld together. Closure mechanisms, then, can stabilize social groups as well as artifacts. A rhetorical closure occurs if a controversy ends not when a neat solution emerges but when a social group perceives that the problem is solved. Closure through redefinition occurs when an artifact stabilized incompletely by one social group is stabilized more completely through association with a larger or more powerful social group. (Pinch and Bijker 1987, 44–46; Bijker 1987; Elzen 1986, 1988). From a systems perspective Hughes (1983) has shown how technological and organizational responses effected an end to "the battle of the systems," the conflict between AC and DC power systems. On a larger scale, stabilization may bring "momentum," a concept Hughes (1987, 76–80; Hughes, 1989) formulated to describe the social processes by which large technological systems shape their own growth and appear to become autonomous. Both systems and momentum appear fruitful in understanding the nature of modern technologies. (Perrow 1984; Mayntz and Hughes 1988; Shrum and Morris 1990; Gökalp 1992) Finally, conflict and controversy underpin the actor-network model developed by Callon, Latour, and Law, who use the metaphor of forming a "black box" to indicate closure. (Law 1987a; Latour 1987; Bowker 1987; Law and Callon, this volume) In a study of Portuguese navigation, Law suggests at least three ways that closure might be achieved—reorienting the lines of force, adapting an artifact to its hostile environment, and using metrication to alter the balance of power—and he even extends the concept to technology itself. "Technology," writes Law (1987b, 128), "simultaneously associates and dissociates, and the heterogeneous engineering of the Portuguese was designed to handle natural and social forces indifferently and to associate these forces in an appropriate form of closure."

Closure must be seen as a contingent achievement of actors and not a necessary outcome of controversies.[3] If achieved, closure implies more than temporary consensus; it is how facts and artifacts gain their "hardness" and solidity. As a social process, closure may frequently involve the creating or restructuring of power relationships. Accordingly, this concept should not be seen as being in opposition to change but rather as facilitating the order that makes change possible. It is only by fixing some elements in place that actors may complete their aim of building systems or constructing networks (Law and Callon, this volume). Indeed, closure may obscure alternatives and hence appear to render the particular artifact, system, or network as necessary or logical.[4] It is precisely because closure can impart direction and momentum that actors battle energetically to achieve closure on terms favorable to themselves.

These themes inform this study of the genesis and resolution of two key controversies in the development of American steelmaking. The first controversy involved a dispute concerning priority for the most important steelmaking process of the latter nineteenth century. In this dispute allies of the English inventor Henry Bessemer clashed with allies of the American ironmaster William Kelly. Closure of this controversy produced a secure legal and patent framework that promoted explosive growth in the industry and thereby engendered a second controversy, which centered on a dispute between two groups that favored rival classification methods to delimit "steel" from "iron." Generally, this essay offers a framework to examine realms typically labeled "scientific," "technological," and "industrial"—the vacillating of the industrial economy, the building of technological systems, and the classifying of new entities—and thus suggests a flexible way to probe the seamless web of history. In analyzing these controversies and the processes of closure that terminated them, this chapter aims to show more than that technologists and scientists argued about steel and then resolved their disputes. Nor is the argument merely that closure facilitated rapid quantitative growth in the American steel industry. The processes of closure, as we shall see, set the boundaries for "steel" and shaped the style and structure of the emerging steel industry. Although this analysis will soon involve powerful industrial combinations, transatlantic negotiations, and railroads that spanned the continent, we must begin with a single ironmaster working in the wilds of Kentucky.

Controversy through Similar Inventions

Like so many technological systems, the Bessemer steelmaking process has a patent controversy at its core. Like other cases, this controversy pits a well-established and powerful entrepreneur against a poorly funded independent inventor. Similarly, the legal arrangements that structured the subsequent development of the technology rest on somewhat shaky documentary evidence. Needless to say, the conflicting claims as to who really invented the process have never been entirely resolved.[5] Hailed by partisans as the true inventor of the "so-called" Bessemer process, William Kelly successfully claimed priority in 1857 when Henry Bessemer applied for an American patent. But in the end Bessemer, or more precisely his allies, reclaimed the upper hand in an 1866 agreement that brought effective closure to the controversy. A review of each inventor's work suggests how this could be so.

Kelly seemed an unlikely candidate to alter the technology of ironmaking. To be sure, his native state of Pennsylvania drew on ready supplies of iron ore, trees, and coal to lead the nation's iron production, and his native city of Pittsburgh drew attention as an emerging iron-rolling center. Kelly entered the iron trade indirectly, however. While he was in Nashville, Tennessee, on business for his wholesale dry goods firm, he fell in love and followed the young woman down the Cumberland River to her native Eddyville, Kentucky. Here in 1846 he bought into the iron trade. Kelly spotted a furnace and forge and interested his brother John in it. Leaving the dry goods firm to a relative, the two brothers took charge of the furnace, forge, and 14,000 acres of land.

The purchase seemed promising at first. Their property offered plenty of trees for making charcoal and it seemed covered with red hematite ore, with a reserve underground. After smelting the ore in a blast furnace about twelve feet in height, which yielded crude "pig iron," the Kellys refined the blast-furnace product with two finery fires. In the finery process about 1,500 pounds of pig iron was sandwiched between layers of charcoal, the furnace was ignited, and then the fire was fanned by a blast of air. The fire removed impurities from the molten crude iron, a process that elevated the iron's melting temperature. When the purified iron stiffened into a pasty ball, one of ten forgemen hammered it into a bar of "wrought iron" perhaps four inches square and several feet long.[6] The finery process consumed copious amounts of timber as charcoal, however, and the next year the brothers relocated closer to timber supply, seven miles from Eddyville at Suwannee. About this same time the surface ore ran out and the Kellys shifted to the underground reserve. The surface and underground ores looked similar to the Kellys' untrained eyes, as neither brother had training or experience in mineralogy, but the underground ore proved vexing. It contained a black flint that refused to burn, and the Kellys were forced to seek another ore supply. Suddenly, with timber and ore in short supply and transport costs prohibitive, the specter of bankruptcy loomed, painfully typical of small forges of the period. This situation probably prompted William Kelly and his brother to experiment.

What experiments, then, did Kelly actually do? Testimony of the Kellys' workmen collected in 1857 for the patent interference case against Bessemer suggests some clues. These documents confirm that Kelly experimented in 1847 and again in 1851 with blowing air into molten iron and that he claimed to have invented a new process, but the testimony denies that he achieved a workable process.[7]

In any event, Kelly did not develop his process. Following his experiments his process lay dormant and unpatented. Only in 1856 did Bessemer's announcement prompt Kelly to act. He secured space to experiment at the Cambria Iron Works in Johnstown, Pennsylvania. It was on the strength of these three bouts of experimentation that Kelly successfully claimed priority over Bessemer's patent in the 1857 patent dispute (see figure 4.1). Two years later, in 1859, he devised a blast pipe to force air into molten metal. He then tried forcing "carbonic acid gas" through the molten metal. Finally, he built a converter that blew air into molten metal through a pipe just below the surface, but this experiment sparked a fire that destroyed the building. "I never heard even a tradition of a perfect conversion made in this vessel," recalled one steelman (Hunt 1876, 210). Kelly's effort stopped except for one final round of experiments in 1862 (see figure 4.2). Eventually he returned to Kentucky to manufacture axes until his death in 1888.

Kelly illustrates a backyard approach to invention. With little capital, a haphazard approach, and little knowledge of the available scientific writings (mostly in German and Swedish), Kelly contrasts markedly with better-known inventors who would follow. Reliable evidence supports neither Kelly's claim to have realized "a process for making malleable iron and steel" nor the sobriquet of "a scholar in metallurgy." (Swank 1892, 399; Boucher 1908, 34) One may grant that Kelly invented a process for air "boiling"—if one defines invention as the act of conceiving an idea and limits "boiling" to its contemporaneous meaning.[8] Available evidence does not, however, suggest that Kelly reduced this idea to practice. Nonetheless, Kelly and his unrealized process influenced developments to come through his patent, which Americans would need to make the new steel.

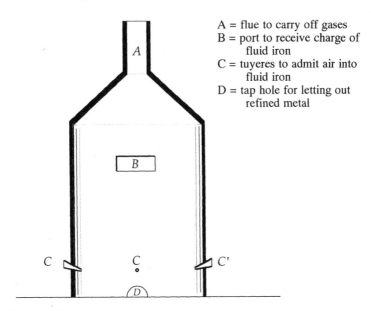

A = flue to carry off gases
B = port to receive charge of
        fluid iron
C = tuyeres to admit air into
        fluid iron
D = tap hole for letting out
        refined metal

Figure 4.1
Kelly's vertical converter from patent (1857). Kelly's patent specification persuaded American authorities to disallow Henry Bessemer's application for an American patent. The patent drawing, from which this is taken, is believed to be the only extant illustration. Compare with figure 4.3. Source: After patent number 17628 (June 23, 1857).

William Kelly and Henry Bessemer could hardly differ more as inventors. Kelly labored in the backwoods of Kentucky, verging on bankruptcy, without patenting, innocent of how to promote an invention. Bessemer by contrast came to his iron experiments as a seasoned inventor; he aggressively publicized his work and labored tirelessly to develop his inventions. Bessemer also cultivated fruitful relationships with influential persons and institutions that supported his inventive activities.

From the beginning Bessemer proved an opportunist. He needed to look no further than to his father, an inventor himself who had fled Robespierre's Paris for England, where he used his experience at the French mint and membership in the French Academy of Sciences to secure employment at the English mint. Coming to London at age seventeen, Henry made a living by executing artistic castings, printing specialty items, and making dies. As early as 1836 his experiments in electrometallurgy attracted Andrew Ure's attention. Bessemer made his first notable invention for the Inland Revenue Office. Its revenues had been cut by the reuse of old stamps on official

Figure 4.2
Kelly's tilting converter at Cambria Iron (1861–62). Kelly used this tilting
converter at the Cambria Iron Company, Johnstown, Pennsylvania, during a
fourth round of experimentation. Compare with figure 4.4. Source: Courtesy of
National Museum of American History, Smithsonian Institution.

documents. To combat this fraud, Bessemer devised a dated stamp that prevented reuse and reportedly brought an additional £100,000 per year to the treasury.[9] Two further episodes illustrate his ability to court prominent patrons. Windsor Castle tendered the first order for his machine-made embossed velvet. (Bessemer claimed that his sister had served as embroideress to Princess Victoria.)

Another of his machines made bronze powder. Its commercial success, which brought Bessemer a small fortune, began with an order from the Coalbrookdale Iron Company, the industrial empire founded by Abraham Darby. Indeed, income from powdered bronze came to support inventive activity at Baxter House, Bessemer's experimental factory and laboratory in the St. Pancras district of London. Proving himself a versatile mechanical inventor, Bessemer patented new methods for casting and setting type; manufacturing paints, oils, varnishes, sugar, and plate glass; constructing railway carriages, centrifugal pumps, and projectiles; and ventilating coal mines. Bessemer began his experiments on iron with 34 patents; he would collect 117 patents in all.[10]

Working out the details of a new projectile design led Bessemer to experiment with iron. Spurred by the Crimean War, Bessemer devised a way of imparting spin to projectiles shot from a smooth-bore gun. When the British War Office rejected his idea, Bessemer took it to France. In 1854, after a demonstration of the revolving projectile at Vincennes, the officer in charge opined to Bessemer that he mistrusted firing heavy thirty-pound shot from the available twelve-pounder cast iron guns and reportedly asked, "Could any guns be made to stand such heavy projectiles?" This conversation, by Bessemer's account (1905, 134–142, quote 135), set him to work. He returned immediately to Baxter House and within three weeks applied for the first of a series of patents concerning iron and steel.

Bessemer claimed his inexperience in ironmaking as an asset, since he felt less constrained by previous practices. Assisted at Baxter House by his brother-in-law, Bessemer began by melting various mixtures of pig iron, blister steel, and scrap in a reverberatory furnace, producing enough metal to cast a model gun.[11] Bessemer's experiments soon took an unexpected twist, as he related (1905, 142):

Some pieces of pig iron on one side of the bath attracted my attention by remaining unmelted in the great heat of the furnace, and I turned on a little more air through the fire-bridge with the intention of increasing the combustion. On again opening the furnace door, after an interval of half an hour, these two pieces of pig still remained unfused. I then took an iron bar, with the intention of pushing them into the bath, when I discovered that they were merely thin shells of decarburised iron...showing that atmospheric air alone was capable of wholly decarburising grey pig iron, and converting it into malleable iron without puddling or any other manipulation. Thus a new direction was given to my thoughts, and after due deliberation I became convinced that if air could be brought into contact with a sufficiently extensive surface of molten crude iron, it would rapidly convert it into malleable iron.

He showed a creative moment by noting the anomaly of the shell of iron. In working out this insight he invented a process that reshaped the iron and steel industry.[12] The malleable iron Bessemer sought resembled the wrought iron of the Kelly brothers' finery furnaces; however, as we soon will see, that this wrought iron had not been worked or "wrought" bedeviled efforts to classify it unambiguously.

Bessemer next built a crucible with a blowpipe stuck into its center. Into the crucible he poured about ten pounds of molten pig iron and, after thirty minutes of blowing air into the metal, found that the crude iron had become malleable iron. This experiment proved air could decarburize pig iron, yet the crucible remained encased in a furnace that consumed fuel. Then, dispensing with the furnace entirely, he built a four-foot-tall, open-mouthed cylinder with openings, or *tuyères*, to blow air into the metal from the bottom. As Bessemer related (1905, 144), at first the blast bubbled quietly through the seven hundredweight of molten pig iron in the vessel, with the opening emitting some sparks and hot gases. Suddenly, after ten minutes, white flame burst forth. "Then followed a succession of mild explosions, throwing molten slags and splashes of metal high up into the air, the apparatus becoming a veritable volcano in a state of active eruption." Unable to approach the out-of-control converter, Bessemer and his assistants could only watch until the eruption quieted after ten minutes more. On tapping the converter and casting an ingot, Bessemer identified the metal as wholly decarburized malleable iron—the product he desired.[13]

In the following weeks Bessemer sought to tame the violent blow. To deflect the flame he placed a cast iron grate above the opening, but the white-hot blast melted it. He then tried lessening the blast by decreasing the number of *tuyères*, their diameter, or the pressure of the air. These changes reduced the temperature, in one case so far as to leave an entire converter full of solid iron. Resigning himself to a violent blow, Bessemer sought to contain it by building a converter with an upper chamber whose side port vented the hot gases. With this vertical, fixed, two-chamber converter Bessemer achieved success (see figure 4.3).

To announce his process Bessemer chose the high road. He showed his converter to George Rennie, president of Section G (Mechanical Sciences) of the British Association for the Advancement of Science (BAAS). Rennie was a good choice because from the 1830s this section of the BAAS had frequently discussed iron technology, including James Neilson's hot blast furnace and Robert Mallet's work on sea-water corrosion. Rennie was impressed and secured Bessemer to give an account at the upcoming BAAS meeting in Cheltenham. On August 13, 1856, Bessemer presented "The Manufacture of Malleable Iron and Steel without Fuel." If this seemingly absurd concept produced snickers before the talk, few remained afterward. Remarked James Nasmyth, this is "a true British nugget." (Bessemer 1905, figure 93). The next day's report in *The Times* attracted the attention of the Dowlais Iron Company's chemist, who prompted the prominent Welsh firm to erect a small trial converter; two weeks later Dowlais purchased the first license for £20,000. (Jones 1988, 51) In early September Bessemer staged an exhibition at Baxter House for "some seventy or eighty of the most eminent persons connected with the manufacture of iron" (Jeans 1884, 44). Sales of licenses netted £27,000 within a month.

Bessemer had announced his process as if it were trouble-free, but it manifestly was not. When Dowlais attempted to roll the first steel rail in 1858, the ingot—one of Bessemer's own—cracked open while still hot; many others found the new metal too brittle to roll or forge into useful shapes. The faltering of commercial ventures eventually forced Bessemer to buy back his first licenses and return some £32,500. At the same time, Bessemer lost the American patent dispute to ironmaster Kelly. What saved Bessemer was his experience with earlier ventures in velvet and bronze powder, plus the money they had realized. Having retained full control of the patents, Bessemer made plans to construct a works in Sheffield. In 1858 he and his business partner formed a partnership with the Messrs. Galloway, the boilermakers and engineers of Manchester who supplied Bessemer with machinery. He also worked to devise a converter lining that withstood the white-hot metal and to locate supplies of pig iron that could be converted into a workable steel (though it was not understood until the early 1860s that phosphorus was the element responsible for making the steel brittle). By 1858 Bessemer switched to a tiltable converter—a major innovation (see figure 4.4). The stationary vertical converter required the blast to be on continuously, from

Figure 4.3
Bessemer's vertical converter at St. Pancras, London (1856). After experimenting for eighteen months, Bessemer went public in August 1856 with his process for converting crude pig iron into malleable iron or steel. Compare with figure 4.1.

Figure 4.4
Bessemer's tilting converter at Sheffield (1858). Bessemer constructed a profitable steel works in Sheffield using the new tilting design, after his first patent licensees failed to make workable steel with his process. Compare with figure 4.2. Source: Henry Bessemer, *An Autobiography* (London: Offices of *Engineering*, 1905), plates XVII and XVIII.

before the crude metal entered the converter until the converted metal left; the blast could cool the converted metal if it had to be held in the converter. With the converter tilted hydraulically or by gears, the blast needed to be on only during the blow itself. During both filling and pouring the vessel tilted so the metal did not cover the *tuyères*. Due to expenses of development, the Sheffield works lost £1,800 in its first two years of operation (1858–59), but was profitable thereafter. (Lord 1945–47, 165–167) With appropriate (low-phosphorus) pig iron, a better vessel lining, and the tilting converter, Bessemer had brought his process to maturity.

An outline of technological controversies emerges from the genesis and aftermath of the 1857 patent interference. First, the myth of "simultaneity" of invention must be laid to rest. The experiments of Bessemer and Kelly, as described above, resembled one another only superficially. Indeed, it is not the inventions themselves but the historical actors, due to their perception of commercial rivalry and conflict, that precipitated the controversy. Second, irregularities and ambiguities in the procedure for granting patents as well as judging interferences render official decisions of little value in determining priority for an invention, which is of little matter here. Nonetheless, the considerable energy devoted by historical actors in prosecuting patent battles makes them an attractive area to investigate. Ample and sometimes voluminous documentary evidence frequently remains to assist the analyst in opening up the "black box" of technology.[14] And finally, patent court decisions may or may not bring closure. In principle controversies can always be reopened. Whether a particular effort does terminate a controversy is a contingent matter that requires investigation and analysis.

Closure through Organizational Innovation

The activities of the two inventors did not complete the drama but rather set the stage for conflict on a broader level. The clash between Bessemer and Kelly during the 1857 patent interference appeared decisively to favor Kelly. But within a decade the first attempts to build converters on American soil engendered a second controversy, involving not the two inventors but their respective allies as well. As this section shows, it was through organizational innovation that the two rival parties achieved closure and thereby shaped the style and structure of the emerging steel industry. In this effort Alexander L. Holley's contacts with many members of the industrial community—but especially with railroad executives and managers—would play a critical role in the shape and definition of Bessemer technology in America.

Holley's brilliant and prolific engineering career—he designed all but one of the first generation of Bessemer plants—followed on his immersion in railroading. Between graduating from Brown University in 1853 and 1860, he designed locomotives for George Corliss and the New Jersey Locomotive Works, published two books based on first-hand knowledge of European railways, and served as technical editor for the *American Railway Review*. The next year, working for Edwin Stevens, the longtime treasurer and manager of the Camden and Amboy Railroad, Holley traveled in Europe to gather information about shipbuilding, armor plate, and armament. He visited Bessemer's works in Sheffield, where he observed the new process as a prospective licensee. "The Bessemer process of making steel," Holley observed in his widely acclaimed *Treatise on Ordnance and Armor* (1865, 104), based on this trip, "promises to ameliorate the whole subject of ordnance and engineering construction in general, both as to quality and cost."

Holley returned to the United States in 1863 and found several) parties interested in Bessemer technology. The assistant secretary of the navy had recently urged the prominent engineer John Ericsson to establish a Bessemer plant. Ironmaster John F. Winslow and banker John A. Griswold had sponsored Ericsson's effort to build the ironclad *Monitor*, and when Holley consulted Ericsson he put the young man in touch with them. These two businessmen from Troy, New York, provided steadfast support for the steelmaking venture.

With this support Holley completed acquisition of the Bessemer technology. To secure an American license for Bessemer's patents Holley crossed the Atlantic in the summer of 1863, his fourth time, and arranged to pay Bessemer £10,000. Winslow, Griswold, and Holley set up a new partnership and began planning a Bessemer plant at Troy. The following year Holley was once again in England, as he put it (Dredge 1898, 939), "to finish my education in the Bessemer process." Notwithstanding the solid legal grounding of the Bessemer process in England, the American licensees soon found themselves facing a complicated lawsuit.

A group headed by Eber Brock Ward had secured rights to William Kelly's patents, and this group now challenged Holley and the Troy entrepreneurs. A businessman active across the Midwest, Ward held shares in transportation, minerals, banking, and iron ventures. In the mid-1850s he had organized the Eureka Iron Company (where south of Detroit a new air-blowing steelworks would be built) and the North Chicago Rolling Mill Company (which would roll the new plant's steel into rails). In 1861, in search of information about the Bessemer process, Ward's advisor Zoheth S. Durfee traveled to England. There he negotiated with Bessemer but failed to sign an agreement on patent rights. Durfee's subsequent travels took him to several other Bessemer works in England, Sweden, and France, including the James Jackson works at Saint-Seurin from which he secured a manager for the new works south of Detroit. To finance this venture, Ward and Durfee formed the Kelly Pneumatic Process Company in May 1863 with capital from Pittsburgh ironmen, including James Park, Jr.

As the two rival American groups built their new steel plants a legal impasse loomed. "Litigation of a formidable character was imminent," as Holley put it.[15] As noted above, Kelly successfully claimed an American patent interference against Bessemer, and Ward controlled Kelly's patents, which covered the concept of blowing air through molten metal. Ward also controlled the American rights to Robert Mushet's patents for treating steel with manganese, which improved the metal's mechanical properties. On the other hand, the Troy group possessed the American rights to Bessemer's patents, which included the tilting converter. The most advanced steelmaking practice used a combination of all these patents. Neither group could construct a state-of-the-art steelworks without infringing on the other.

In 1866 an out-of-court settlement resolved this legal block by exploiting a new organizational form. The precise course of negotiation remains unknown, but the outcome became clear enough. The patents were pooled and split: the Troy group received 70 percent of the proceeds from licensing fees; the Ward group, 30 percent. To administer the patent pool the two groups set up "The Trustees of the Pneumatic or Bessemer Process of Making Iron and Steel." This organization—subsequently reorganized as the Pneumatic Steel Association, the Bessemer Steel Association, and the Bessemer Steel Company—did not operate plants directly but provided a means for licensing the pooled patents, collecting royalties, and dividing the proceeds.[16]

The founding of the Bessemer Association terminated controversy, and in this case closure was coextensive with the emergence of a legal basis for the new industry. For the immediate future the Bessemer Association assured that no protracted legal battle would check growth; from 1866 to 1877 it licensed eleven plants. Commanding the only legal route to Bessemer technology, the Bessemer Association as an institution served the entrepreneurial function of presiding over change (Misa 1985; Eisinger 1988; Andersen and Collett 1989). By creating a legal mechanism to share the needed patents (and a financial arrangement to apportion proceeds from the licenses), the Bessemer Association constituted itself as an "obligatory point of passage" for those who would manufacture the new steel (Latour 1987, Law and Callon, this volume).

Not only did this form of closure promote change, it also shaped subsequent development. First of all, after 1877 the Bessemer Association effectively prevented others from legally acquiring the technology it controlled, which included several key patents of Alexander Holley, by restricting the number of licensees. Moreover, by cooperating with the railroads to set production quotas and prices, the Bessemer Association bypassed the invisible hand of a competitive market and imposed order on the steel rail trade. Into the next century, various cartels alternated with market mechanisms to set prices on steel rails. And as late as 1908, railroad business composed 60 percent of the total output of the Association of Steel Manufacturers.[17] Steelmakers and railroads thus shifted from "anonymous regulation," or a market-based form of interaction, to "private regulation" with agreements among producers and consumers (Jacobs 1988; Lundvall 1988).

Controversy through Market Control

Those who introduced the Bessemer process did more than found a bulk steelmaking industry. They also upset the traditional craft-based control of iron and steelmaking and initiated a chain of events that shaped the new science of steel. At Bessemer plants especially, craft-oriented metallurgists were pushed aside when managers hired employees who had no experience in the iron trade (and hence no "bad" habits to unlearn). Conflicts sometimes flared up between the craft metallurgists and the university-trained chemists who took their place (Misa 1987, 30–31). But significant conflict also emerged *within* the community of science-oriented metallurgists and professional engineers. Most vexing, as Holley's epigraph to this chapter suggests, was the problematic nature of "steel." The controversy that erupted in the mid-1870s concerning this issue not only points to significant links among the scientific, technological, and business realms, but also permits empirical inquiry into the "communal strategies" that underlay the emergence of two rival classifications to delimit "steel" from "iron" (Barnes 1982, 27–31, 76–80, 101–114).

Since the 1850s, mechanical, physical, and chemical methods had been available to define iron and steel. Data from mechanical tests were the easiest of the three to obtain. To measure mechanical strength, testers subjected samples to torsional, tensile, and compressive forces. The Army's huge Watertown Arsenal testing facility became the leading center for mechanical testing, and the mechanical engineer Robert H. Thurston stepped forth as the chief spokesman for this approach. For its advocates, "iron" had certain mechanical properties, "steel" had others. Physical tests such as density were a second possibility. Desirable mechanical properties could be correlated with the density of reference pieces of metal and unknown samples classified by density alone. Most advocates of density tests believed that metal became stronger as density increased (being free from holes or inclusions), but some felt there existed an optimum density that signaled maximum strength.

Mechanical or physical standards were inadequate, however, to those concerned with the iron and steel industry on a regional or national level. Density, for example, could be useful to compare batches of pig iron made in the same blast furnace from the same iron ore, but density did not reliably distinguish iron or steel samples from different regions. Similarly, mechanical testing could identify a particular piece of metal as being desirable, but mechanical testing could not identify how it was made or how other desirable pieces of metal just like it could be made. That mechanical and physical tests were insufficient to yield general rules about the properties of iron and steel became evident to Alexander Holley. "'A bar of steel' is, in the present state of the art," complained Holley (1873b, 117), "a vastly less definite expression than 'a piece of chalk.'"

From his broad overview Holley could point out the weakness of mechanical and physical testing. He noted that although engineers and machinists often complained they could not regularly obtain a certain quality of steel, thousands of tons of steel had been made that were entirely suitable. To standardize the steel trade Holley pointed to chemical properties. "In order that engineers may know what to specify, and that manufacturers may know not only what to make, but how to compound and temper it," he continued, "the leading ingredients of each grade of steel must be known." Chemistry, for Holley, was central to the new process. "The difficulties of the Bessemer manufacture," he once observed, "were not chiefly mechanical," but stemmed from "a chemical stumbling-block." And describing a converter's blast he exclaimed, "What a conflict of the elements is going on in that vast laboratory!" Holley (1873b, 117–119; Holley 1873a, 1; Holley 1868, 19).

The adoption of chemical methods also reflected the increasing geographical size of the steel market. With the transcontinental rail boom of the early 1870s, the larger railroads became ventures of national scope. Rails from central Pennsylvania could be laid down as far away as California. Especially for the railroads, larger markets created the problem of guaranteeing quality at a distance. As the principal buyers of Bessemer steel, railroads selected and defined its essential properties. Charles B. Dudley, the Pennsylvania Railroad's Yale-trained chief chemist, persistently advocated chemical specifications for steel rails, though his effort to do so was not without controversy. New and larger markets probably required some standard to guarantee quality at a distance, but the specific form of the standard owed much to the railroads.[18]

Metallurgists soon found that their knowledge could be used not only to standardize markets but to manipulate them. In the mid-1870s an acrimonious debate erupted concerning a deceptively simple question: What is steel? The debate, though conducted in scientific language, had immense commercial implications. Recast, the question really became who would be designated to make the valuable commodity "steel" and who would be left making "iron." Participants openly articulated their respective commercial and professional interests. (More accurately, they identified the interests of their opponents.) Whereas high-temperature steelmakers supported the "fusion" classification, low-temperature steelmakers and university metallurgists affirmed the "carbon" classification.[19] The genesis of this controversy helps account for its form.

The timing and scope of the controversy stemmed from instability in the iron and steel trade. To begin, the depression following the panic of 1873 (itself triggered by overextended speculation in railroad bonds) pushed down prices and overall economic activity for several years. Because railroads consumed more than half the total iron produced and imported, the halting of railroad construction produced a severe slump in the iron and steel trade. In early 1874 rail mills were running at less than one-third capacity, with some 21,000 rail-mill workers thrown out of full-time employment. Not until late in the decade did the iron and steel industry recover.

An ongoing switch from iron to steel also led to instability. Railroads were shifting their orders from iron to steel rails, and, as the price gap between steel and iron narrowed, analysts forecasted a dim future for iron rails. Production statistics confirmed these fears. Iron rail output peaked in 1872 at 809,000 tons, then fell steadily to a mere 9,000 tons in 1889. Total iron output fell after the 1873 panic, but recovered by the end of the decade. In contrast, steel production grew continuously and vigorously. In 1870 total steel output stood at 69,000 tons; by 1880 it topped 1,200,000 tons. At this time Bessemer steel composed 86 percent of total steel produced, and rail mills consumed 83 percent of total Bessemer steel (Temin 1964, 270, 274–275, 278, 284–285).

Finally, even steelmakers were experiencing unsettling shifts. In Pittsburgh, Andrew Carnegie snapped up several undervalued steelworks following the 1873 panic. When his Edgar Thomson Works, designed by Holley to mass-produce Bessemer steel rails, began production in 1875 the Bessemer Association nearly collapsed. Concentrating on rails, Carnegie's firm challenged the Bessemer Association's control over this sector, and steel rail prices fell to $42 per ton in 1878. To producers who had sold steel rails in 1873 at $120 per ton, these were confusing times indeed. (Steel rail prices inched up to $67 per ton in 1880, then fell again to around $30 per ton by mid-decade.)

Iron- and steelmakers reacted to these shocks in two ways. The industry first attempted to reorganize along oligopolistic lines, as the coal industry had successfully done, to cushion the downward spiral of prices. Then the makers of Bessemer and other high-temperature steels moved to alter the specifications delimiting steel from iron. Because, for example in 1875, "steel" rails sold for $69 per ton whereas "iron" rails sold for $48 per ton, this move sparked a scientific dispute with large commercial consequences.

The traditional method used to distinguish iron from steel relied on carbon content. A critical variable, carbon in small amounts imparted resilience, strength, and most importantly the capacity for being hardened upon sudden cooling, or quenching, from high temperatures. "Wrought iron" contained essentially no carbon; "steel" contained from 0.2 to 1.0 percent carbon; and "cast iron" contained two percent or more carbon. Steel could be hardened by quenching; wrought iron could not.

The spread of the Bessemer process made possible an alternative classification. Whereas traditional methods for making wrought iron (such as Kelly's finery furnaces as well as the process of puddling) yielded a pasty, semisolid mass, the Bessemer converter completely melted or "fused" its products. Metal that had been fused was free from the cinders, slag, carbon flecks, and other inclusions that traditionally characterized wrought iron. According to one account, the resulting "homogeneous" product had qualities that were "universally recognized" if not "readily described." Kicking off the debate, Holley articulated the fusion classification: "Steel is an alloy of iron which is *cast while in a fluid state* into a malleable ingot." By the fusion classification, if the metal had been completely melted—regardless of its carbon content—it became "steel"; if not it remained "iron." [20]

Advocates of the carbon classification rallied behind a young metallurgist, Henry M. Howe. The son of a Boston doctor, Howe had attended Harvard College and the newly founded Massachusetts Institute of Technology. After completing his education in 1871, he gained steelmaking experience by working at the Bessemer works in Troy, serving as superintendent of Joliet Iron and Steel's new Bessemer works in Chicago, and working at the Blair Iron and Steel Works in Pittsburgh. Howe's arguments in the debate implicitly supported the low-temperature iron and steel manufacturers whose products the fusion classification would reclassify as "iron." Howe also upheld the metallurgical tradition that maintained that hardness and resiliency defined steel. In this respect, as well as attempting to keep metallurgy from being shackled to raw economic interests, Howe might be seen to represent the scientific as opposed to the engineering viewpoint. Yet after the debate had quieted, Howe designed and built two metallurgical works for the Orford Nickel and Copper Company—an activity for which Holley was admired as an engineer—and he later served as vice president of a specialty steel-manufacturing firm.

Howe advanced his case in the *Engineering and Mining Journal*. He advocated correlating the mechanical properties traditionally associated with steel—resilience and ability to be hardened—with a sample's carbon content. Carbon content would be an index: "steel" would have the same carbon content (0.2 percent–1.0 percent) as did reference samples of steel. He then critiqued the fusion classification. He noted it had become "fashionable" to label as "steel" all products of the Bessemer converter and open hearth, without regard for mechanical properties. Now, he commented, "cultivated and intelligent engineers" claimed that "the distinction between wrought iron and steel should be based on homogeneousness and freedom from slag, and that hardness, tensile strength, resilience, and the power of hardening have nothing to do with it." Howe suggested how this "confusion" arose. When iron from a Bessemer converter or open hearth furnace was cast, the resulting ingots had not been worked or "wrought" and could not be "wrought iron." Since these ingots looked and felt like steel, some believed the easiest way was "to call the whole product steel, and not bother about mechanical tests, or split hairs about physical properties." The same reasoning, he observed, "would justify a jeweler in selling brass as gold or strass [flint glass] as gems." Howe (1875, 258) explained,

It is possible that some manufacturers, being human, were influenced by the consideration that steel was vaguely associated in the minds of the public with superiority and was in general higher priced than wrought iron, to sell that part of their product as steel which a strict adherence to the then recognized distinction between steel and wrought iron would have compelled them to call wrought iron.[21]

Holley's sharp rebuttal launched the metallurgical community into a full-scale controversy. The American Institute of Mining Engineers, founded in 1871, served as the primary arena of dispute. That the debate, which soon involved pointed personal attacks, stayed within the professional and trade journals testifies to the professional tactic of maintaining a conspiracy of silence against outsiders.[22] In answer to the rhetorical question "What is Steel?" Holley (1875, 138–140) remarked curtly, "The general usage of engineers, manufacturers, and merchants, is gradually, but surely, fixing the answer to this question." He disputed Howe's arguments point by point, contending that whereas the fusion classification was already—or nearly—in place, the carbon classification was "arbitrarily devised" and "must bear the demerit . . . of upsetting existing order and development."

Holley gained support from industrialists connected to high-temperature steelmaking processes (Bessemer, crucible, open hearth). James Park, Jr., an early Pittsburgh investor in the original Ward-Kelly company, attacked Howe and supported the fusion classification. Another fusion advocate was William Metcalf, a prominent crucible steelmaker. After graduating in 1858 from Rensselaer Polytechnic Institute, he returned to his native Pittsburgh as assistant manager and draftsman at the Fort Pitt Foundry, rising within a year to general superintendent. In the late 1860s he helped organize the firm of Miller, Metcalf, and Parkin, which owned the Crescent Steel Works. As managing director, he specialized in fine crucible steels until 1895, when the Crucible Steel Company bought out his firm. In various ways Park, Metcalf, and Holley were each committed to high-temperature steelmaking.

No high-temperature steelmakers stood on the other side. Those rallying behind Howe and the carbon classification included Thomas Egleston, head of the School of Mines at Columbia College,

Frederick Prime, professor of metallurgy at Lafayette College, and Benjamin W. Frazier, professor of metallurgy at Lehigh University, as well as Frank Firmstone, superintendent of the Glendon Iron Works, and Eckley B. Coxe, a prominent anthracite mining engineer. John B. Pearse became an unexpected ally. As a chemist for Pennsylvania Steel (a creation of the Pennsylvania Railroad) since 1868 and its manager since 1870, Pearse "ought" to have supported the fusion classification. But in June 1874 he resigned his steelworks position and became commissioner and secretary of the Pennsylvania state geological survey; in October 1875 he attacked Holley and supported the carbon classification. A year later he became general manager of the South Boston Iron Company. Advocates of the carbon classification shared at least one formal characteristic: "steel" did not appear in their affiliations.

The carbon advocates soon identified the interest behind the fusion classification. Prime (1875, 332) pointed to Holley, then president of the American Institute of Mining Engineers, who "belongs to a group composed of himself . . . and many manufacturers of Bessemer and open-hearth steel, who propose to overthrow the definition I have given as the current one. With energy worthy of a better cause . . . he gives his definition, pronounces *it* to be the current one, and claims that 'several high metallurgical authorities and clever writers have of late proposed to disturb this natural and somewhat settled nomenclature.' (!)" Howe (1876, 516) expanded Prime's argument that commercial interests motivated this gambit:

Certain mechanical engineers and manufacturers, many or most of whom were pecuniarily interested in attaching the name steel to the new products, because that name was associated in the mind of the public with superiority, have called these new products steel, in full face of the fact that they had none of the essential qualities of steel and all of the essential qualities of wrought iron.

Howe tabulated the varying results of the two classifications for standard iron and steel products (table 4.1). Note the difference: the fusion classification redefined the three low-temperature steels (blister, puddled, shear) as "iron," whereas all products of high-temperature processes (Bessemer, Martin [open hearth], crucible) became "steel."

The fusion advocates did their part also to identify the interests behind the carbon classification. Holley and Metcalf portrayed the carbon advocates not simply as scientists but as elitists and autocrats. Metcalf (1876, 357) complained that "the few, the men of science"

Table 4.1
Varying result of classifying iron and steel products

By carbon classification	
Cannot harden—wrought iron	*Can harden—steel*
Puddled iron	BLISTER STEEL
Bloomary iron	PUDDLED STEEL
Malleable castings	SHEAR STEEL
Bessemer iron	BESSEMER STEEL
Martin iron	MARTIN STEEL
Crucible iron	CRUCIBLE STEEL
By fusion classification	
Has not been fused—wrought iron	*Has been fused—steel*
BLISTER STEEL	Bessemer iron
PUDDLED STEEL	Martin iron
SHEAR STEEL	Crucible iron
Puddled iron	BESSEMER STEEL
Bloomary iron	MARTIN STEEL
Malleable Castings	CRUCIBLE STEEL

Source: Henry M. Howe, "The Nomenclature of Iron," *American Institute of Mining Engineers Transactions* 5 (1876): 517–518.

were arbitrarily enforcing an "ancient" meaning of steel, and he chafed at their assumption of authority and superiority. "The names of new materials and processes," added Holley (1875, 147), "are not fixed by the arbitrary edicts of philosophers." Yet again, however, this was no simple division between science and engineering. Holley and Metcalf both claimed the mantle of science and, in fact, argued that their classification was *more* scientific than that of the "high metallurgical authorities and clever writers."

Logic alone does not unravel these debates. Both sides claimed priority for their classification. Both sides maintained that the opposing classification was arbitrary or confusing. And both classifications had technical merit. The ability of "steel" to be hardened, on which the carbon advocates focused, was a property of real significance. Similarly, the fused "steels" had important properties, such as freedom from slag and other inclusions, that unfused "steels" did not possess. Finally, despite their both containing similarly low percentages of carbon, "wrought iron" (unfused) and "mild steel" (fused) were two entirely different products (Bealer 1969, 44–45, 146).

Instead, as the disputants readily identified, behind the debates were varying commercial and professional interests in conflict.

Holley and other high-temperature steelmakers supported the fusion classification: it defined all products of their processes as the higher-priced "steel." Howe and the low-temperature steelmakers attempted to retain the carbon classification: it preserved their professional integrity as well as their markets. The outcome of this controversy would determine the contours of ferrous metallurgy for the rest of the century.

Closure through Social Process

Undesirable consequences awaited the metallurgical community if this controversy were not brought rapidly to an end. In the abstract, two rival classifications could coexist, but several practical problems emerged that challenged the metallurgical community's legitimacy as experts who dealt in reliable knowledge and objective facts. Import duties were one such problem. In May 1878, after eighteen months of lobbying, William Sellers—the machine-tool magnate who had recently reorganized the Midvale Steel company—and others persuaded the secretary of the treasury to reclassify imported Siemens-Martin metal, a fused product which had entered the country under the (lower) iron tariff to the detriment of American steel manufacturers, as "steel." Thereafter, perhaps to the chagrin of the scientific metallurgists, "collectors of customs" were "to make the proper classification."[23] The controversy spilled over the Atlantic in another way. American and European metallurgists initiated a joint effort to develop a unified nomenclature of iron and steel, but differences between the national contexts as well as linguistic shadings between English, German, and French paralyzed this effort. Finally, noted Metcalf (1880, 551), there was "a heavy suit pending in the United States courts, turning upon the question whether steel is steel or iron." For these reasons resolving the controversy grew ever more urgent. To anticipate, the American iron and steel community adopted the fusion classification but retained chemical methods. No less than the controversy itself, closure involved economic, technological, and sociological factors. It may justly be seen as a social process.

If the panic of 1873 had sparked the debate, the improved economy of the early 1880s helped extinguish it. By 1880 manufacturers had seen orders surpass even predepression levels, and they were not as nervous as in 1873. Moreover, by 1880 the price gap between iron and steel rails had virtually closed as Carnegie's Edgar Thomson works, along with a half dozen other modern mills, flooded the market with mass-produced steel rails. Until the next classification debate—which followed the 1893 panic—metallurgists maintained consensus on the boundaries of "steel."

The establishment of the fusion classification owed much to the technological needs of the railroads, still the largest consumers of steel. Railroads had found that rails rolled from fused metal, even with carbon content similar to wrought iron, were less likely to crack open than rails made from unfused metal. The superintendent of Pennsylvania Steel (Pearse 1872: 163) noted of steel rails that "their homogeneity is their distinguishing characteristic." By supporting the fusion classification, railroads ensured that the metal that best filled their specific technological needs would be uniformly available.[24] And railroad financiers no less than railroad managers and steelmill owners appreciated the advantages that this wondrous metal conferred. One financial analyst (Swann 1887, 35–37) identified the adopting of steel rails as a strategy to inflate the value of a railroad's stock for speculative purposes.

By entering management, metallurgical chemists themselves contributed to the momentum of the Bessemer process and chemical metallurgy. As experts in process control, they quickly rose to be managers, as several biographies illustrate. Robert W. Hunt took a course in analytical chemistry from the Philadelphia chemists Booth and Garrett, his only formal education. At Cambria's Bessemer works at Johnstown, Pennsylvania, he established the first chemical laboratory associated with an iron and steel firm in America. In 1867 he rolled the first commercial order for steel rails, delivered to the Pennsylvania Railroad. Thereafter he held a succession of managerial posts, retiring as the head of a consulting firm in Chicago. Booth and Garrett's teaching laboratory also trained John B. Pearse, who entered the laboratory with a B. A. from Yale in 1861, stayed two years, then studied at the Freiburg School of Mines for another year. As noted above, he began work as the chemist for the Pennsylvania Steel works and within three years was promoted to general manager. Another Pennsylvania Steel chemist, Edgar C. Felton, who joined the firm in 1880 after graduating from Harvard, rose through the ranks to become the firm's president in 1896. The rise of chemists into management could be high indeed. James Gayley served for three years as chemist to the Crane Iron Company in Pennsylvania's Lehigh Valley before moving through a series of managerial posts beginning in the 1880s. He capped his career in 1901 when he became first vice president of the United States Steel Corporation.

Finally, what of the issues that divided the advocates of the "carbon" and "fusion" classifications? If bitterly contested, the differences between Holley and Howe were mediated by practice. Holley consistently advocated the use of chemical composition to standardize the *varieties* of steel; the fusion classification served only to delimit steel from wrought iron. While Howe also advocated chemical methods to classify steels, he failed to maintain carbon content as the single method to delimit steel from wrought iron. It was in this regard that the fusion classification triumphed.

By 1880 debate on the method to classify steel was terminated, and the once-problematic categories "iron" and "steel" were relatively stable. "Steel" emerged as a metal having been fully melted, "iron" was incompletely melted, and both were subject to chemical standards. Economic, technological, and sociological factors together conspired to shape a metallurgy based on the fusion classification (for delimiting "steel" from "iron") with a chemical component (for classifying the varieties of "steel").

Conclusion

The processes of closure that terminated each of these two controversies brought "hardness" or obduracy to a contested field and thus structured change. For the patent dispute between the allies of William Kelly and Henry Bessemer, the closure achieved by the founding of the Bessemer Association helped shape the style and structure of the emerging steel industry. For the dispute between rival advocates of the carbon and fusion classifications, closure terminated debate, but did not result in a new institution. Yet each instance of closure influenced developments to come, and each was durable beyond the short term. The Bessemer Association, working in concert with the leading railroads, nurtured the Bessemer steel rail sector in the 1860s and imparted to it characteristics (nonmarket price-setting mechanisms and the switch to "private regulation" among producers and consumers) that endured into the twentieth century. Following the termination of debate around 1880, the consensus on "steel" was strong enough to prevent a flare-up of public debate, except during 1893 when that year's financial panic prompted a small classification controversy.

Although the mechanisms that effected order and stability were clear enough in the Bessemer steel rail industry, it is less clear how the consensus on "steel" was maintained. To begin, no technical commission or professional body successfully pronounced a definition of steel as authoritative. Only after 1900 did metallurgical textbook authors grant canonical status to the fusion classification.[25] Even then, the definition of steel was "in a shockingly bad condition," according to Bradley Stoughton, Henry Howe's assistant at Columbia University. Into the 1920s, even the bible of the steel industry, compiled by the Carnegie Steel Company, wrestled inconclusively with a textbook definition of "steel."[26]

It appears that "steel" owed its stability less to written authority than to the daily practice of thousands of steelworkers and managers. Howe (1891, 1) maintained that the fusion classification would not have become dominant if "the little band [of his fellow carbon advocates], which stoutly opposed the introduction of the present anomaly and confusion into our nomenclature, [could] have resisted the momentum of an incipient custom as successfully as they silenced the arguments of their opponents." The "incipient custom" was, of course, that of the railroads, Bessemer steelmakers, and their allies such as Alexander Holley. It is in this complex of powerful interests that the forces that fundamentally shaped the science and technology of steelmaking are to be found. It is no accident that Holley was a staunch advocate of modern railroading, the Bessemer process, and the fusion classification—in that order.

Because the railroads and Bessemer steelmakers achieved closure on terms favorable to themselves, they imparted a decidedly conservative direction or momentum to the steel industry. The Bessemer Association, alongside other industrial cartels, railroad pools, and trade associations popular (and still legal) in the late nineteenth century, constituted an "amalgamation of vested interests" (Bijker, this volume) among established ironmakers, railroads, and entrepreneurs. Not only did the established social network stabilize the new technical network, but the technical reinforced the social. Iron rolling mills built eleven of America's first thirteen Bessemer steel plants; the nation's largest railroad built another; its officers invested in Carnegie's Edgar Thomson mill, the only new mill of the lot (Misa 1987, 35–38). This sociotechnical network exercised considerable power not merely because of its social connections (exercised through the Bessemer Association); it maintained and extended its power through promoting the new Bessemer technology, which yielded a commanding position in the marketplace. Stabilizing the technological order meant stabilizing the social order (Todd 1987; Latour 1987; Misa 1988b).

In the steel industry, a conservative pattern of technological change emerged by the 1870s and became well entrenched before

1901, when J. Pierpont Morgan formed the United States Steel Corporation around Carnegie's holdings. In this and other core industries, Morgan's leadership forestalled the "creative destruction" of the old by the new once posited by Schumpeter as a natural pattern of industrial growth (Douglas 1987; Galambos and Pratt 1988, 5–37). Corporate industrial research efforts of the early twentieth century reinforced the Morgan paradigm of conservative change (Noble 1977; Reich 1985; Hounshell and Smith 1988; Hughes 1989, 138–183, 459–461). Reconciling the technological determinist's rhetoric of radical, technology-driven change in society with the reality of conservative change managed by vested interests has remained problematic ever since (Misa 1988a and 1992).

The history of steelmaking is typically portrayed as the onrush of new technologies or the building of massive industrial empires. In this chapter we have added the role of scientific knowledge to this heady brew and attempted systematically to relate the scientific, technological, and economic realms. We found no realm to be privileged. Scientific classifications as well as technological processes were subject to disputes that were deeply technical. Yet an analysis focused on the technical logic of the disputants' positions does not explain these disputes or their resolution. For Bessemer and Kelly no less than for Holley and Howe, considerations of technical merit merged with personal or professional gain. Analysts must appreciate how participants articulated objective positions imbedded in a web of commercial or professional interests. The resolution of a dispute was not the product of philosophical ratiocination or the accumulation of facts, for not all objective positions were equal. Rather, closure reflected the interplay of interests, aspirations, and power. The themes of controversy and closure encourage detailed analysis of the social alongside the technical. Such an approach also reveals the social processes that serve to construct and maintain the objectivity of facts and artifacts.

Notes

Acknowledgments: For comments and helpful questions I would like to thank Nil Disco, Thomas Hughes, Robert Kohler, Henrika Kuklick, Judith McGaw, Maurice Richter, Edmund Todd, and Steve Usselman, as well as the referees of and several contributors to this volume. For bibliographic assistance I am indebted to Clemens Moser of the Eisenbibliothek.

1. For a discussion of explaining and understanding "objective culture," see Kuklick 1983.

2. For controversy and closure as constructive social processes shaping the emergence of facts and artifacts, see Holton 1978, Morrell and Thackray 1981, chap. 8, Collins 1985, Rudwick 1985, chaps. 13 and 16, Secord 1986, Bijker, Hughes, and Pinch 1987, Engelhardt and Caplan 1987, Latour 1987, Hull 1988, Gooding, Pinch, and Schaffer 1989, and Todd 1989.

3. See the "endless" controversies discussed by Dean (1979), MacKenzie and Barnes (1979), Pinch (1986), Segerstråle (1986).

4. The perceived "necessity" or "logic" of technological development is one of two types of technological determinism discussed in Misa (1988a and forthcoming). The focus is here on change, but investigation of the social processes that yield stability or continuity should not be deemed an ahistorical task. On stability, see Maier (1975, chap. 8; Maier 1987); on technology and social continuity, see Todd (1987).

5. Kelly's partisans claimed that Bessemer stole Kelly's secret by visiting Kelly incognito; see Boucher 1924 and Boucher 1908, 39–41; cf. "Dedication of Tablet Recalls Bessemer Patent Controversy," *Iron Trade Review* 71 (10 Oct. 1922), 1064. This contention is all but dismissed by McHugh (1980, 121–123). Conversely, Bessemer and his partisans claimed Kelly appropriated Bessemer's ideas by copying his patent specification; see Bessemer 1896, 413. Most accounts of the Bessemer-Kelly dispute rely on Swank (1892, 395–400), which reproduces Kelly's own account. For full documentation, as well as greater detail, see Misa 1987, chap. 1.

6. The crude iron tapped from the blast furnace solidified in molds that formed bars or "pigs"—hence, in English, "pig iron" (in German, *Roheisen*; in French, *fonte brute*). In modern terms, this product contained 4 percent or more carbon and varying amounts of other impurities. If pig iron was simply remelted and cast into pots, stove plates, or other items for which a hard but brittle metal was suitable, it became "cast iron" (*Gußeisen, fonte*). As noted in the text, crude iron could be purified by burning out the impurities and then "working" the metal by hammer or rolling mill—hence, "wrought iron" (*Schmiedeeisen, fer forgé*). As the German and French terms suggest, this was a soft and malleable metal well suited for being bent, shaped, or welded at the blacksmith's forge; in modern terms wrought iron contained essentially no carbon. "Steel" (*Stahl, acier*) was the tough yet malleable metal traditionally understood to be intermediate between wrought iron and cast iron (or pig iron) and capable of hardening when cooled quickly, or quenched, from, say, a red heat. I have used "iron and steel" to refer generically to the industry, "iron" to refer to the making of wrought iron or cast iron, and "steel industry" to refer principally to the Bessemer process. (In the latter third of the nineteenth century there were two other fundamental steelmaking processes: the Siemens-Martin process, or open hearth furnace, accounted for small but increasing output, in both absolute and relative terms; while crucible steel accounted for small and decreasing output, in relative terms. See Misa 1987, 363–364.)

7. State of Kentucky, Lyon County, Interference. *William Kelly* v *Henry Bessemer*, Apr. 13, 1857, I: 28 Patent Suit *Kelly* v *Bessemer* 1857; American Iron and Steel Institute Papers (Acc. 1631); Hagley Museum and Library, Greenville, Delaware.

8. In the parlance of mid–nineteenth-century metallurgy, "boiling" meant a variant of "puddling," the traditional process used to make wrought iron. Neither featured a Bessemer-like blast of air *through* molten iron.

9. Bessemer claimed £100,000; also noted in *Dictionary of National Biography*, s.v., "Sir Henry Bessemer," XXII: 186; Tweedale 1984.

10. Given the paucity of reliable documents, historians must interpret Bessemer's boastful and possibly unreliable *Autobiography* (published posthumously in 1905); notable for comparing Bessemer's account with primary documents is Birch (1963–1964); see also Jones 1988. On Bessemer as a mechanical inventor, see Bessemer 1905, 4–137, 329–332, and Lange 1913, 2–5, 33–8.

11. Blister steel resulted from baking bars of iron that contained essentially no carbon in furnaces filled with charcoal; the process, known as cementation, required as long as a week before the desired amount of carbon in the charcoal diffused into the iron bars, producing blisters on their surface; blister steel melted in clay crucibles became the "crucible steel" that made Sheffield famous. See Tweedale 1987, chap. 2.

12. Others had treated molten iron with blasts of air or steam. For discussion of experiments similar to Bessemer's, see Wertime 1961, 284–287.

13. In modern terms, the oxygen in the air and the silicon and carbon in the metal combined to produce a violent combustion; such a chemical understanding was, of course, yet to emerge.

14. A detailed, behind-the-scenes analysis of a patent dispute over steel armor is Misa 1987, chap. 3. The American patent system is discussed in Hindle 1981, Noble 1977, chap. 6, Reich 1985, chap. 9, and Post 1976. The English patent system is discussed in Dutton 1984, Hewish 1987, and Macleod 1988.

15. Holley quoted in "The Invention of the Bessemer Process," *Engineering* 6 (27 Mar. 1896): 414.

16. See the account statement Bessemer Steel Association to Maryland Steel 29 Jan. 1897 17: 5 Maryland Steel General Correspondence 1897; Maryland Steel Company Papers of Frederick W. Wood (Acc. 884); Hagley Museum and Library.

17. H. S. Snyder to A. Johnston, 26 Mar. 1908, RG 109: 3; Bethlehem Steel Papers (Acc. 1699); Hagley Museum and Library. From May 1901 through 1906—while the price for nails, beams, bars, billets, and pig iron fluctuated on market demand—steel rails remained set at $28 per gross ton. See "Fluctuations in the Prices of Crude and Finished Iron and Steel from January 1, 1898, to January 1, 1907," *Iron Age* supplement (10 Jan. 1907); Temin 1964, 173–182.

18. By 1880 chemical specification of rails was standard in America, according to a leading European authority on rail specifications (Sandberg 1880, 205): "The specification for rails, since the introduction of the use of steel, is becoming almost overdone—in America chemically, with the stipulation of only one certain chemical composition in the rails." "In America the control and inspection of the quality of steel is overdone in another direction, viz., by chemical analysis, so that steel, even for rails, is now nearly always chemically analyzed and the composition stipulated in contracts" (Sandberg 1881, 406). Dudley's work is dissected in Usselman 1985, esp. 295–299, 300, 317.

19. For contemporaneous comment, see Williams 1890, chap. 9.

20. Holley 1872, 252–254 [original emphasis], Holley 1873a, 2–3, Howe 1876, 515. C. W. Siemens (1868, 284), the inventor of another high-temperature steelmaking process (open hearth), had earlier announced the fusion classification.

21. For Howe (1875, 258) steel was "a compound or alloy of iron whose modulus of resilience can be rendered, by proper mechanical treatment, as great as that of a compound of 99.7 per cent. iron with 0.3 per cent. carbon can be by tempering."

22. For the conspiracy of silence as a professional tactic, see Johnson 1972, 1977. In the ensuing controversy pitting carbon and fusion advocates against each other, the AIME's secretary, Rossiter W. Raymond, provided the only consistent voice for compromise.

23. *Engineering and Mining Journal* 25 (8 June 1878): 396. "Everybody connected with the steel trade knows how irregular are the duty rates, and desires more clearness and simplicity," noted Greiner (1877, 138).

24. Holley (1875, 142) observed that the Pennsylvania Railroad "specifies 0.35 carbon steel for its rails, meaning by 'steel,' that it shall be homogeneous or cast." Howe (1875, 259) admitted as much: "A Bessemer rail . . . having no welds to yield to the incessant pounding, usually lasts till it is actually worn out by abrasion. Hence, railway managers do not care very much about the degree of carburization of rails said to be steel, provided they are absolutely weldless, and a steel rail has come to mean with them a weldless rail instead of a hard rail. They are, in general, willing to receive all the products of the Bessemer converter as steel, provided they are not too brittle. Were their pleasure alone to be consulted, freedom from welds might be the most convenient ground for the classification of iron."

25. This is clear in a sample of twenty-seven English-language metallurgical textbooks drawn from the Eisenbibliothek, Schaffhausen, Switzerland, and the University of Chicago's Crerar Library. Before 1880 the carbon classification dominated (100 percent); during 1880–99 the sample split evenly between carbon and fusion classifications (several writers gave both); by 1900–09 the fusion classification (60 percent) triumphed over the carbon classification (27 percent) and the emerging microstructural classification of metallography (13 percent).

26. Stoughton (1934, 44) observed, "For fifty years a struggle waged between scientific metallurgists who wanted to call low-carbon iron 'wrought iron,' and manufacturers who wanted to call it 'steel' if it had low carbon and no slag [i.e., had been fused]. The situation has finally been clarified greatly by the practical obsolescence of the cementation process" (which yielded a non-fused steel). Earlier, Stoughton (1908, 7) offered a definition of steel that simply lumped the carbon and fusion classifications together: "Iron which is malleable at least in some one range of temperature, and in addition is either (a) cast into an initially malleable mass; or (b) is capable of hardening greatly by sudden cooling; or (c) is both so cast and so capable of hardening." The Carnegie definition did little better (Camp and Francis 1925, 254–255): "Before beginning the study of modern methods of producing steel, it is desirable to decide the question as to what steel is. Owing to the many varieties of iron now classed as steel, a concise and wholly satisfactory definition is well nigh impossible. Attempts have been made to restrict the usage of the term, but without success, because, in defining any term, the name must be taken as it is used." The Carnegie definition of steel was as a residual category: "*Steel* is the term applied to all refined ferrous products not included under the classes above [pig iron and cast iron, malleable cast iron, wrought iron]. It is distinguished from pig iron by being malleable at temperatures below its melting point, from malleable iron by the fact that it is initially malleable without treatment subsequent to being cast, and from wrought iron by the circumstance of its manufacture [i.e., having been fused]."

5

Closing the Ranks: Definition and Stabilization of Radioactive Wastes in the U.S. Atomic Energy Commission, 1945–1960

Adri de la Bruhèze

In most historical analyses of the radioactive waste problem, the issue suddenly emerges as a social, scientific, and political problem in the early 1970s, at the time that the U.S. Atomic Energy Commission (AEC) announced that it would build an underground repository for high-level radioactive waste in an abandoned salt mine near Lyons, Kansas. Though there is some truth in this, it does not explain how the problem of high-level radioactive waste could suddenly leap to prominence, after seemingly having been neglected for twenty-five years. Neither does it question the ahistorical assumption that there had always existed, and only could exist, a single, "natural," and unequivocal meaning of radioactive waste.

This chapter has two interrelated objectives. First, I want to show how radioactive waste was socially shaped as an ontological entity to become the problem that we recognize today. Second, I want to present an analytical framework to deal with the heterogeneous character of technological development in the setting of a large organization. It will be argued that the social shaping of radioactive waste and radioactive waste technology within an organizational structure provided the institutional actors with different opportunities and constraints for shaping a problem and a technology in conformity with their ways of defining and approaching problems. I will do this by showing that the radioactive waste problem was socially shaped by the interaction of a series of actors who defined problems and their possible solutions in ways that were heavily influenced by their education, training, and institutional affiliation (Pinch and Bijker 1984; Elzen 1986; Vergragt 1988). I will suggest that the actions and solutions proposed by actors were shaped by the way in which they articulated problems, while "action" and "practices" simultaneously structured the social process of problem definition. The reciprocal relationship between problem definition and practices, or social shaping and social impact of technology, are described in more detail by Bijker (1987, this volume) with his concept of "technological frame." In particular I will trace the way in which relevant actors tried to preserve, expand, or spread their way of seeing and treating radioactive waste in the large governmental AEC. Because the AEC was exclusively charged with the entire management, development, regulation, and promotion of nuclear technology, including the handling of its radioactive waste, it constituted the central organization in the development of nuclear technology in the United States in the period 1945–1960. In order to promote their views on waste, relevant actors in the AEC used strategies to limit the number of participants. In particular, to stabilize a set of social and technical relations in the AEC, these actors sought to bureaucratize and technologize social and political problems, even though these did not necessarily relate to radioactive waste. I will show how these processes developed by focusing on the way in which relevant actors interacted in a series of interconnected bureaucratic fora, where a variety of resources were mobilized, gained, and lost, and where certain views about waste and its treatment became connected, gaining strength, while others weakened and were lost.

By the early 1960s, the outcome of this process was a generally shared view about the central problems and how they should be solved. In the AEC, this stabilization and closure was experienced as a "natural" fact, instead of a social and bureaucratic construction. This made the AEC think that radioactive wastes could be safely dealt with [and that there was no need to actively maintain the stabilized problem and solution definition]. This attitude toward the internally stabilized views on radioactive waste and its management was one of the main reasons why actors who had been excluded from participation between 1945 and 1960 were unexpectedly able to redefine radioactive waste as an important social problem in the early 1970s.

Settings of Radioactive Waste

The problem of radioactive waste and its treatment has been deeply embedded in two important and interdependent settings. First, there is the context of nuclear technology and the specific way in which its development has been organized. Second, there is the context of the daily operational activities and practices of local laboratories. I will deal with each in turn.

The development of nuclear technology in the United States was a government undertaking delegated to the AEC. Military and civilian technology could not be separated from each other, at least not until 1960. At first nuclear technology served primarily military purposes: producing plutonium in reactors or "piles" for atomic fission bombs, or developing atomic power reactors for the propulsion of ships and aircraft. Thus, most radioactive waste was, in the first instance, the responsibility of the military. As a result of the prevailing secrecy about all military nuclear activities, the existence, quantity,and composition of these wastes was a closely kept secret.

Although nuclear technology was then primarily a military technology, its civilian aspects were loudly proclaimed and politicized during the cold war. Its manyfold future peaceful applications were advertised as the showpiece of U.S. technological superiority, which would profoundly affect and change all spheres of human life. This process occurred before useful applications actually existed. In fact, in the 1950s and 1960s the continuous effort to build and maintain public support for the government's military and civilian atomic policies depended on such promises about the future peacetime uses of atomic energy; potentially disturbing information about the lethal effects of radioactivity, and thus the existence and problems of radioactive waste, did not fit such utopian images. (Boyer 1985; Del Sesto 1987). Thus, unlike the peaceful future applications of nuclear technology, the existence of radioactive waste in the AEC was consciously hidden from the outside world. In short, the problem was depoliticized and bureaucratized.

In the context of daily operational practice—the second setting for the management of radioactive waste—the problem was seen differently. Thus the various AEC laboratories created all kinds of liquid, gaseous, and solid radioactive fission products and materials. These materials were defined as "ashes, poisons, scrap, or waste" with various degrees of radioactivity. Something was done with those wastes, but there was no general rule or procedure, nor was it felt necessary to develop preferred ways of handling waste.

Bureaucratized Radioactive Waste

In the late 1940s the AEC contained five headquarter divisions: reactor development/engineering, research, production, biology and medicine, and military application. The production division, charged with the production of plutonium and other fissionable materials for atomic weapons, supervised most of the national laboratories and production sites (see figure 5.1).

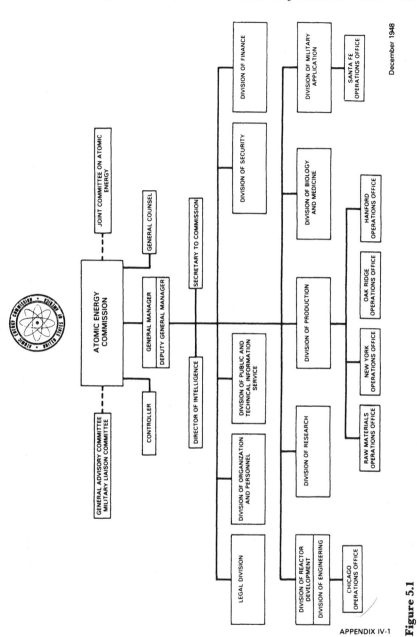

Figure 5.1
U.S. Atomic Energy Commission organizational chart.

APPENDIX IV-1

18

Between 1945 and 1960, no headquarters division was exclusively responsible for radioactive waste management at the national laboratories. Rather, waste management was just one aspect of the overall operation of the laboratories, which were responsible to a range of headquarter divisions. As a result, and in combination with the complex and often conflicting organizational responsibilities and the practiced policy of "operational flexibility" (Metlay 1985; Hewlett, 1978), the local laboratories acted rather autonomously with respect to the definition, characterization, treatment, storage, and disposal of radioactive waste. Different kinds of high- and low-level radioactive wastes were distinguished in a range of different ways.

Low-level solid radioactive wastes were buried on-site in the ground or off-site at sea, whereas liquid and gaseous radioactive wastes were diluted and dispersed. The existence of various low-level waste-handling practices was legitimated by saying that the laboratories had different tasks, were involved in different processes, and generated different kinds of waste, and that different environmental conditions permitted different dispersal and disposal practices.

Liquid high-level wastes were stored in various kinds of underground tanks. These wastes were commonly defined as the residues remaining after the chemical reprocessing of irradiated uranium fuel elements from military or "production-purpose" reactors. Most were produced and stored at the production sites of Hanford (Washington) and, from 1954 onward Savannah River (South Carolina). Smaller amounts were produced at the national laboratories such as the National Reactor Testing Station (NRTS-Idaho), Oak Ridge National Laboratory (ORNL-Tennessee), Argonne National Laboratory (ANL-Illinois), and Brookhaven National Laboratory (BNL-New York). However, there was, as might be expected, substantial variation between the different locations not only in the practices but also in the interpretations of radioactive wastes.

Relevant Actors and Their Definitions of Waste

In the early 1950s, different actors with different definitions of radioactive waste could be distinguished. To differentiate between the different positions of these actors and their views, I will use two dimensions: global/local and internal/external. The two dimensions, when combined, generate four cells: global-internal and local-internal, which refer to groups inside the AEC, and global-external and local-external, which refer to groups outside the AEC (see figure 5.2).

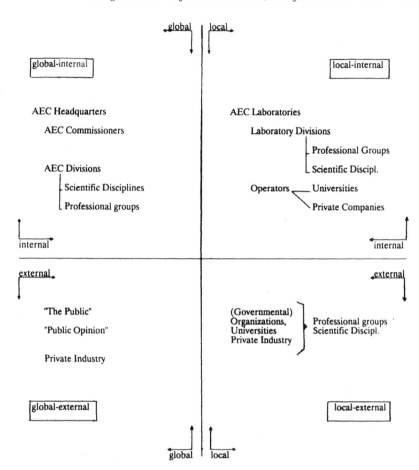

Figure 5.2
Positioning of relevant actors.

AEC Headquarters (Global-Internal)

For the *AEC commissioners*, AEC's top-level decision makers, radioactive waste constituted a "normal" and non-urgent problem. They were confident that technically and economically feasible solutions would be found for the treatment, storage, and disposal of radioactive waste. In reaction to growing public awareness and fear about radiation from bombs and other AEC operations at the end of the 1940s, the AEC started an atomic energy information campaign to educate the public and allay possible fears. The campaign was aimed at building support for AEC policies and acclimating the public to the reality of living with radiation (Boyer 1985). As an important feature of this campaign, a public report on radioactive waste was issued in 1949.[1] The publication or the report was delayed by more than a year because of military resistance, the existence of different views on radioactive waste in the AEC, and disagreements about the environmental and public health hazards of radioactive waste disposal practices.

For the *Division of Military Applicatication* and other military people in the AEC, such as members of the Military Liaison Committee, who coordinated Department of Defense's objectives and AEC programs, radioactive waste was not a problem at all, but instead a useful source material for all kinds of weapons and other military applications. Within this perspective, various fission product applications were considered and investigated: poisons for radiological warfare, sources for the irradiation of food, tracer isotopes (Hacker 1987, 46–49; Ridenour 1950) Defined in this way, fission by-products were no waste at all. As a consequence, the military wanted to keep information about this source material as secret as possible and successfully resisted publication of the first versions of the 1949 waste report because "it would be of substantial assistance to a competitor nation because of its authenticity and association with our processes, rates of production, and recovery operation."[2]

For the *Division of Production*, radioactive wastes constituted the unimportant and unattractive by-product of the plutonium production process. Members of the division's chemical engineering group responsible for waste management at the production sites usually saw liquid high-level radioactive wastes as chemical wastes, and their radioactive toxicity as chemical toxicity. Therefore, they fully supported the local treatment practices that used techniques applied to chemical waste. The chemical and nuclear engineers in the division thought that long-term tank storage of liquid high-level wastes, as well as other waste-treatment practices, would be adequate and safe.

Furthermore, they considered the wastes stored in tanks as an abundant source of usable radioisotopes, which like plutonium could be chemically extracted from the liquids.

The Division of Biology and Medicine, which funded research programs of health physicists, biologists, physicians, and radiologists, saw radioactive wastes in terms of human health risks. According to this division, radioactive wastes had to be handled carefully, and highly radioactive isotopes had to be extracted from the waste to make environmental disposal and dispersal safer for AEC personnel and the public. This extraction also would yield useful radioisotopes for therapeutic use, though the division was very concerned about the increasing practice of disposing medical radioisotopes into the sewers.

The Division of Reactor Development/Engineering was responsible for the management of radioactive wastes from new and future reactor types. Therefore it was responsible for assessing and developing technologies for handling future wastes. In this division, where chemical engineers, mechanical engineers, nuclear engineers, and physicists collaborated, radioactive waste was generally seen as a technical and economic problem that could (and should) be solved in an economic way to avoid hampering future reactor development and power generation. The division delegated its work on waste to its *Sanitary Engineering Branch* (SEB). Though I will discuss the work of this branch in more detail below, here it is important to note that the sanitary engineers, trained to handle large amounts of sewage and industrial wastes, defined radioactive waste in terms of public health and environmental safety. This view could have conflicted with the problem definition of the Division of Reactor Development, with its primary concern with cost reduction; the sanitary engineers, however, adapted their position to include the economic requirement in their problem definition.

With respect to existing waste-handling practices, the Sanitary Engineering Branch considered tank storage of liquid high-level waste as both a public health hazard and an expensive, unwieldy practice, so they dismissed it as the ultimate solution. Tank storage was only acceptable for the sanitary engineers as an interim solution prior to the permanent and safe disposal of radioactive waste. As possible permanent disposal solutions they considered the concentration of waste into the smallest practicable package, followed by geological and sea disposal. The sanitary engineers were concerned about the existing ("operational") treatment and handling practices of low-level waste and instead investigated the feasibility of applying sanitary engineering methods for water purification to confine radioactive and chemical toxicity to permissable levels prior to environmental disposal and dispersal.

The divisions were generally not very interested in the local practices of the laboratories; the views of the latter about radioactive waste became important as soon as there was a debate within the AEC or an externally directed action like the publication of a report. For the Sanitary Engineering Branch, however, the situation was different. It was their responsibility to make sure that waste was handled properly, and from their background they brought very specific ideas about what should be done.

AEC Laboratories (Local-Internal)

As previously described, AEC laboratories defined and handled radioactive wastes in different ways. However, they generally agreed that radioactive waste was not an important or urgent problem, and it could be technically solved. Moreover—largely due to investments they had made in their operational practices—the laboratories showed an interest in maintaining and expanding their own waste-handling practices and waste-research programs. The people who considered existing radioactive waste management to be a serious problem were found among the sanitary engineers, health physicists, physicians, and biologists who worked at programs that were sometimes funded by the Division of Biology and Medicine, The Division of Research and the Sanitary Engineering Branch.

Private Industry (Global/Local-Internal/External)

In its daily operational activities the AEC heavily relied on its primary contractors, mainly universities and large industrial corporations. Although private industry generally felt that waste management was not attractive as a profit-making undertaking, some chemical industries were interested in the commercial possibilities of recovering radioisotopes from the stored high-level waste. Overall, however, the general attitude of industry toward waste management was that it was not attractive because of the existence of too many unresolved technical and administrative questions. The primary responsibility for the development of ultimate waste-disposal methods should lie, the industry argued, with the AEC. This attitude was strongly influenced by the refusal of insurance companies to insure radioactive waste-disposal activities (Dorsett 1950; USAEC 1957a).

The Public (Global-External)

Immediately after the Second World War, nuclear technology and especially radioactivity had an ambivalent significance for the American public. Atomic energy and radiation were cheered and feared at the same time, permitting fantasies, expectations, fears, and speculations to grow (Hine 1986; Weart 1988; Boyer 1985; Del Sesto 1987). Radioactivity became the pivotal element in the public perception of American nuclear technology because it linked all nuclear issues, ranging from fallout, reactor safety, and nuclear accidents to nuclear wastes. To allay public anxiety about radioactivity by promoting the peaceful atom, the AEC was forced not only to take public relations seriously but to consider technical safety considerations and waste management as well. It was within this perspective of waste as a sociopolitical problem that the AEC published its 1949 waste report.

In this confusing and complex interplay of tasks, responsibilities, funding, practices, interests, strategies, and different definitions about radioactive waste and its problems, there was no fixed demarcation between local and global spheres of influence with respect to waste-management responsibility in the period 1945–1960. So far, I have described the actors already in place in 1949. However, in 1949 the Sanitary Engineering Branch (SEB) was established, a formation that can be seen as a first step in the demarcation of internal/external and local/global distinctions.

The Sanitary Engineering Branch

The Sanitary Engineering Branch could be called the brainchild of Professor Abel Wolman, Professor of Sanitary Engineering at Johns Hopkins University and a national authority on water resources and public health. While serving in an AEC advisory committee evaluating existing safety and health practices at the local laboratories, Wolman became deeply concerned about existing waste-handling practices. Through his acquaintance with AEC Chairman David E. Lilienthal, Wolman successfully urged the establishment of a small environmental unit in the AEC that had to pave the way for a future division of health protection. The small environmental unit, the Sanitary Engineering Branch, was established within the Division of Reactor Development/engineering and contained until the mid-1950s only two sanitary engineers, Arthur E. Gorman and Joseph A. Lieberman. Gorman was Wolman's personal friend, ally, and confidant, and Lieberman had received his doctorate under Wolman in 1941.

From the moment the Sanitary Engineering Branch was created, Wolman, Gorman, and Lieberman cooperated closely in their attempts to acquire and mobilize all kinds of internal and external resources to increase interest in the environmental aspects of radioactive waste management in the AEC. Wolman for instance became consultant for the AEC, chairman of AEC's Stack Gas Working Group, and a respected member of the AEC's Reactor Safeguard Committee.

> The AEC's full-time environmental unit together with the Reactor Safeguard Committee gave me an opportunity to educate those who, for many good reasons, were either unfamiliar with the environmental issues or rated them low on the totem pole. (Hollander 1981, 400)

Initially the Sanitary Engineering Branch, charged with the research and development of environmental aspects of atomic energy, and, to a lesser extent, the environmental aspects of AEC operations, tried to get agreement on waste definition and characterization and what should be done about it. Furthermore, the branch set up a independent R&D program and funded local R&D work—the only way to get things done at the laboratories. The SEB tried to persuade other divisions to fund and expand this work and to change existing waste-handling practices but, given the lack of direct authority, this had to be done by means of careful persuasion and "working relationships."[3]

The importance of the Sanitary Engineering Branch grew with the passage of the *second atomic energy act* in 1954, which was aimed at the development of commercial nuclear power. AEC's Division of Reactor Development expected that large quantities of radioactive wastes would be generated by commercially built and operated power reactors and reprocessing plants. The Sanitary Engineering Branch therefore had to evaluate and assess technologies for civilian nuclear waste management as a contribution to the development of a strong and internationally competitive American nuclear power industry. The act offered the SEB an additional opportunity to play a major role in future commercial radioactive waste handling and to alter existing military waste-handling practices. The Hess Committee resulted from the SEB's efforts to press its waste management views on the AEC. This committee, and especially its Princeton Conference, was the first forum where actors and their different views on radioactive waste could interact.

The Hess Committee: The SEB's First Attempts at Persuasion

In 1953 the SEB asked Wolman's Sanitary Engineering Department at Johns Hopkins University to study the feasibility of sea and land disposal of future commercial high-level wastes. The university group recommended in 1954 that land disposal be studied further by the Earth Science Division of the U.S. National Academy of Sciences, (NAS) and that sea disposal of highly radioactive wastes was not warranted because of limited oceanographic knowledge. The SEB followed this advice and asked the Earth Science Division conduct a feasibility study. The NAS division created a committee composed of non-AEC geologists and chaired by Professor Harry Hess, a Princeton University geologist (see appendix 2). The newly formed Committee on Waste Disposal was charged with the organization of a conference to generate ideas on geological disposal of high-level radioactive wastes, which was held on September 10–12, 1955, at Princeton. Sixty-five people participated—attended for the most part geologists, but also oil and mining engineers from universities, industry, and government agencies.

Gorman, head of the SEB, set the agenda of the conference:

Almost every year appropriations must be made to build more and larger tanks, but this cannot go on forever. We are looking to this group for more rational schemes directed toward disposal to the ground.... The problem has really two major categories: 1) where and how can we put wastes into the ground economically and under conditions which will not jeopardize the rights of others, especially in populated areas; and 2) what can we do with the large volume of wastes that have been and are yet to be produced at our production plants, particularly those which are being accumulated in underground tanks at the Hanford works in the state of Washington. (USNAS 1957b, 17)

Gorman's associate Lieberman, stressing the long half-life of the existing high-level radioactive wastes stored in tanks, added that "In other words, as presently practiced, tank storage of high-level wastes is not actually disposing these materials" (USNAS 1957b, 35).

Thus, at the start of the conference the Sanitary Engineering Branch made it clear that it was looking at the possibilities of geological disposal for future commercial and existing military high-level radioactive wastes.

The proceedings of the Princeton conference reveal the fascinating process of how a group of people tried to evaluate a solution to a problem embedded in a heterogeneous and in many ways secret technology with an uncertain future. It was at that time unclear which reactor type would be the winner, what kind of nuclear fuel would be chosen, what kind of reprocessing method would be adopted, and whether usable radioisotopes could be extracted from the commercial liquid high-level waste. These interdependent aspects of future commercial nuclear power were important because they would determine the character, composition, and amounts of future commercial high-level radioactive waste. Secrecy about existing high-level radioactive wastes, however, made it difficult to assess future commercial radioactive waste. Nevertheless, the only way to make progress in the assessment of commercial waste appeared to evaluate the efficiency and technical suitability of existing treatment and storage practices used for military waste. In addition, and contrary to the case of military waste, the SEB defined safety and economics as fundamental to the development or a civilian nuclear waste technology and the smooth development of a competitive nuclear power industry. Guided by these considerations, and with a background in geology and mining, the Hess Committee decided to examine methods used by the oil industry to store liquified petroleum gas, petroleum, and oil in subsurface cavities and to dispose oil and brine wastes in geological formations, hoping that these methods would offer solutions for the radioactive waste problem. Distinguishing between deep and near-surface disposal (6000–7000 feet), two subcommittees were set up.

Based upon the experience of its members with geological brine waste disposal, the deep disposal subcommittee concluded that liquid high-level radioactive waste with a greater specific gravity than oil might be pumped into the bottom of a geological structure from which oil had been or was being pumped out higher up. For near-surface disposal of high-level radioactive waste, the other subcommittee recommended geological salt formations. This recommendation sprang from the petroleum geologists' knowledge that there were many salt beds and domes in the United States that could be cheaply acquired. In addition it was known that salt conducted heat and had a high melting point, and that cracks in salt domes tended to seal themselves under pressure.[4]

The subcommittees also discussed the form in which future commercial waste should be disposed. After considering existing solidification programs of the AEC laboratories in terms of cost, safety, and technical feasibility, the near-surface group advocated solidification as a long-term goal. Temporarily, while the technology of solidification was being developed, waste could be disposed of in liquid form in suitable salt formations. By contrast, the deep-disposal group recommended direct disposal of liquid waste and that future nuclear installations be built above suitable geological formations.

The conclusions of this and other studies conducted by the Hess Committee were presented in a report that appeared in April 1957 (USNAS 1957a). The main conclusion was that radioactive waste could be safely disposed of in many but by no means all areas of the United States. The recommendations were presented as an agenda, that is, a gradually realizable, practical approach to the disposal of existing and future commercial high-level radioactive waste. Safe disposal was defined as complete isolation from the biosphere for at least 600 years, the period high-level wastes were considered hazardous. Existing military tank-storage practices were judged to be safe and reasonably economical,[5] but safer and more economical methods of disposal would be needed for future commercial high-level radioactive waste. In the short term, while new systems for waste management were developed, tank storage would be required as an interim solution. The most promising method for the near future was likely to be direct liquid disposal in geological salt formations, rather than solidification which, although better in some ways, posed considerable technical problems.[6] In the long run, however, the emphasis should shift to solidification, according to the committee. In addition, the presence of suitable geological formations for waste disposal should determine the location of all future nuclear installations, especially chemical separation plants, producing large quantities of highly radioactive waste.

This last recommendation met substantial resistance in the AEC because it could be seen as an extension of the controversial recommendations of the AEC's Reactor Safeguards Committee on the crucial issue of reactor siting.[7] This committee had recommended that high—power and potentially dangerous reactors be restricted to more isolated spots, a suggestion resisted by the AEC's General Advisory Committee, the AEC Commissioners, and the Division of Reactor Development, which wanted to accelerate and expand the civilian power reactor program, and which implied siting of reactors near population centers.

The Wolman Committee: Radioactive Waste and Growing Social Concern about Radiation

The Princeton conference was generally seen as a success and this, together with the report of the Hess Committee, strengthened the position of the SEB, in part because the Hess Committee was considered independent and authoritative. However, at the same time AEC attention was drawn to growing public concern about radioactive fallout. When radioactive ashes of the atmospheric hydrogen bomb test explosion "Bravo" on Bikini Island contaminated 28 Americans, 250 islanders, and the 23 crew members of the Japanese fishing vessel *Fukuryu Maru* in March 1954, the health and genetic effects of radioactive fallout became a major scientific and public issue in the United States (Divine 1978; Kopp 1979). Increasing public concern about the rising levels of fallout could be detected in public opinions polls and newspaper coverage (Kraus, Mehling, and El-Assal 1963; Rosi 1965; Gallup 1972); some newspaper articles even linked the dangers of radioactive fallout with the possible radioactive dangers of future peaceful nuclear power (Mazuzan and Walker 1984). This prompted the trustees of the Rockefeller Foundation to ask the U.S. National Academy of Sciences (NAS) to examine the biological and genetic effects of radioactivity. In April 1955 the NAS announced the establishment of a standing Committee on Biological Effects of Atomic Radiation (BEAR), composed of six subcommittees containing more than 100 prominent scientists. One of the subcommittees was the Committee on Disposal and Dispersal of Radioactive Waste, chaired by Professor Abel Wolman.[8]

Wolman selected as members of his committee scientists who were, one way or another, engaged in waste management and who were not overtly hostile to Wolman's sanitary engineering approach (see appendix 2). Among the committee members were Lieberman and Gorman from the SEB and Hess and Theis from the Hess Committee. The Wolman Committee had a broad fact-finding task: it had to arrive at conclusions and suggest topics requiring further research about the safety of radioactive waste management. Unlike the Hess Committee, a committee of and for professionals, with a low public profile, the Wolman Committee was part of a prestigious endeavour to settle an intense public and scientific debate. Well aware of this, the AEC Laboratories had an interest in claiming that their existing waste-handling practices were safe and economical.[9] The internal tension generated by the interplay of disciplinary rapprochement, different organizational interests, and the committee's very public role, made the Wolman Committee another relevant forum where actors with different waste perspectives met and interacted.

The Wolman Committee started its work by examining and assessing its task during a conference in Washington, D.C. on

February 23 and 24, 1956. Unlike the Princeton conference of the Hess Committee, the Wolman conference was a closed meeting without contributions by outside experts. During the conference the committee members defined their task broadly[10] and discussed existing waste-handling practices and possible safe treatment methods in terms of cost, efficiency, and safety. They concluded that the management of existing and future high-level radioactive waste was a solvable problem that should not hamper the development of commercial nuclear power. In addition, representatives of the AEC laboratories stressed the safety of their local programs and practices and opposed another BEAR subcommittee, the Subcommittee on Genetic Effects of Atomic Radiation, which was suggesting that permissable radiation dose levels be reviewed.

Three members of the Wolman Committee wrote to Wolman in April 1956 that new radiation standards were not necessary because "conventional practices in most installations impose a voluntary safety factor. This voluntary self-regulation is not economically crippling, and one may reasonably expect the same order of control in the future atomic power industry." If the permissable radiation doses were decreased, "almost all current instrumentation would have to be discarded," and "many current operations could not be performed," because these would lead to unacceptably higher costs.[11] Taking issue with this, Hess, Theis (Hess Committee), and Gorman (SEB)[12] stressed the feasibility of new methods for geological disposal of existing and future high-level radioactive waste. This tension between different problem definitions and possible solutions became even clearer when the three scientists noted that none of the existing laboratories was in a area suitable for geological disposal. Future nuclear installations should, they argued, be built only at sites with geological formations suitable for the disposal of high-level radioactive waste. Hess was therefore dissatisfied with those parts of the draft summary report of the Wolman Committee that characterized existing storage and disposal practices at the AEC laboratories as nonhazardous.[13]

On June 12, 1956, the parent BEAR Committee published its findings in two forms: a collection of the summary reports of the subcommittees and a further condensed version designed for the public (USNAS 1956a, b). Hess was delighted with the summary report of the Wolman Committee, which recommended both research on the geophysical and geochemical aspects of ultimate waste disposal, and geological site selection criteria for future nuclear installations. In fact, all members of the Wolman Committee could be satisfied because the summary also declared that the waste-disposal practices of the laboratories were safe and economical. The summary report therefore concluded that existing waste-disposal practices were satisfactory and had no harmful effect on the public or the environment, but that environmental consideration had to he given to future commercial radioactive waste generated near populated areas. Not surprisingly, the summary report recommended that the different disposal systems worked at or proposed by the AEC laboratories be further developed to bring them to the point of economic and engineering reality. On fallout, the report briefly stated that it had not been "an environmental contaminant of substantial public health significance" (USNAS 1956a, 107/108), although the problem would need continuous attention if more frequent weapon testing were to occur.

Internal tensions of the Wolman Committee, and objections from the Sanitary Engineering Branch and the Hess Committee about the summary report's remarks on waste-handling practices, were more visible in the public BEAR report. In this report, which was based on the reports of the chairmen and the rapporteurs of the six subcommittees (including Wolman and Lieberman), frequent reference was made to radioactive waste from nuclear power reactors by comparing their possible biological effects to nuclear fallout. The public report explicitly rejected atmospheric dispersal and sea disposal of radioactive waste because of the lack of meteorological and oceanographic knowledge, and implicitly advocated geological disposal of radioactive waste.[14]

After the publication of the BEAR reports, the Wolman Committee continued its activities. Especially Wolman, Lieberman, and Gorman attempted to utilize the committee as a useful forum for over coming and reconciling different interests and views on radioactive waste management. The continuity of the committee was impeded, however, by the tendency of security officers at the AEC laboratories to classify the members' reports. In addition, Wolman had difficulty keeping his group together; meetings were infrequent and poorly attended, and most reports did not appear on schedule. It seemed that the majority of Wolman Committee members were no longer interested in the work of the committee after the completion of its 1956 summary report, which had been published in the BEAR summary report (USNAS 1956a).

Nevertheless, some committee members tried to complete and extend the 1956 summary report by writing a handbook on radioactive waste disposal. In June 1957 this second Wolman report was published (Culler and Mclain 1957). It was not definitive and only an internal report, but it became widely used as a working document in the field and was described as a Wolman Committee report, though it was never considered official by committee members.[15] In this report, attention was paid to existing disposal practices and to the future disposal of commercial high-level radioactive waste. Safe disposal of high-level waste was defined as the "containment" of radioactivity, until then by storage in tanks. However, the latter was treated as a temporary expedient for cooling military and (future) civilian high-level wastes prior to their treatment and permanent disposal. As a permanent solution, the report favored disposal of cooled liquid or solidified wastes in geological salt formations. The report also criticized existing geological disposal practices of low-level waste at Hanford and Oak Ridge National Laboratory, saying they would be unsafe if continued into the future, a view that coincided precisely with that of the Hess Committee. The second report, and especially its criticism of disposal practices, became a matter of controversy within the Wolman Committee itself, and the representatives of the local laboratories did not hide their irritation about this criticism and interference. During a September 1957 meeting, representatives of the laboratories, and especially those of Hanford, attacked the criticism that existing disposal practices had received in the second report and questioned the feasibility of the geologic disposal alternative. The Hanford representatives even explicitly adhered to permanent tank storage. At the end of the meeting, Wolman could only conclude that the report needed fundamental changes before it could be published. However, incompatible views on handling wastes, different interests, extensive written comments, delays in the distribution, and other priorities of the Wolman Committee members delayed publication until December 1959.

Until September 1957, the Wolman Committee played an important role during the fallout debate by providing support for AEC's public declarations and legitimating existing waste-handling practices. However, the committee also constituted a forum for the most important actors, at which opinions could be expressed independent of daily institutional contexts. The second Wolman report was an extreme example of this independently expressed opinion and revealed some of the conflicts that could be generated and made visible in this way.[16]

The Hess and Wolman Committee: Differences and Conciliation

After the Princeton conference and shortly before the 1956 publication of the Wolman report, Gorman and Lieberman and members
of the Wolman Committee, who had initiated, partly organized,
and participated in the Princeton conference, drafted the view of the
Sanitary Engineering Branch on the future disposal of commercial
high-level waste.[17] Strengthening their sanitary-engineering-problem
definition of radioactive waste by linking it with the conclusions
of the Princeton conference and the findings of the Wolman Committee, Gorman and Lieberman tried to bring the SEB's waste
definition onto the agenda of the Division of Reactor Development
and, through this division, onto the agenda of the AEC. This strategy
appeared to be successful; the branch's opinion was accepted by the
Division of Reactor Development and submitted, in March 1956, to
the AEC commissioners as the division's opinion. In their report,
Gorman and Lieberman wrote that the disposal of (military) radioactive waste was under control, but that future commercial high-
level radioactive waste was "a problem whose thread runs through
the entire fabric of nuclear energy operations." According to the
authors, this problem would be formed by the high cost of waste
handling and disposal, the high radioactivity and toxicity of the
waste, its environmental health effects, and the fear that possible
public concern could hamper the development of nuclear power.
Public relations problems were expected to occur when the atomic
industry started to build power reactors close to their markets, that
is, near large cities. "*The public relations problems* in disposal of radioactive wastes to the environment are very important if the industry
is to avoid punitive controls."[18] The growing importance of the
sanitary-engineering-problem definition was stressed when the report
was released by the Division of Reactor Development in May 1956.[19]

Despite or perhaps because of its success, the Hess Committee
sought a continuing advisory role to the AEC shortly after the
Princeton conference. In October 1956 Lieberman, who perceived
the committee as a useful future ally, agreed to this request, envisag-
ing that it might periodically review research and development and
evaluate proposals for future waste disposal. This led the Hess Committee to believe that it would remain the most important advisory
committee on the geological disposal of radioactive waste for the
AEC as a whole, and it sought a clear distinction between its role and
that of the Wolman Committee in order to avoid possible future

conflicts and to define its own influence on AEC waste-disposal policy. In the negotiations between Hess, Wolman, and the executive staff of the National Academy, it was agreed that the Wolman Committee would conduct broad research to locate safety problems in radioactive waste treatment and disposal, and the Hess Committee would advise the AEC on the geological disposal of such waste. The latter committee interpreted this task as a permit to comment and advise on all nuclear developments having geologic aspects and implications. However, despite these different tasks, the work and the courses of the two groups tended to overlap.

The conciliation of the Hess and Wolman Committees became visible in the initial agreement among the Sanitary Engineering Branch, the Hess Committee, and the Wolman Committee, through the second Wolman report, and the *new* geological disposal solution proposed by the Hess Committee. However, as we have seen, this initial agreement was not welcomed everywhere. The AEC laboratories represented in the Wolman Committee rejected the criticism of existing waste-disposal practices at AEC laboratories and questioned the feasibility of the geologic disposal alternative. The laboratories only agreed that the disposal alternatives should be confined to future commercial wastes.

This outcome, and the controversial site-selection criteria defined by the Hess Committee for nuclear installations, made it difficult for the Sanitary Engineering Branch to press its (general) definition of (all) high-level radioactive waste on the AEC commissioners and divisions. This was clear in the report on nuclear waste disposal published in August 1957 by the Division of Reactor Development (USAEC 1957b), in which it indicated its policy for future commercial waste. In this report the disposal practices at Hanford, Oak Ridge, and Savannah River criticized by the Hess and Wolman Committee were characterized as "examples of improved operations which have resulted from the development work" (USAEC 1957b, 26). Proposals for future waste-management policy, based on the work of the Sanitary Engineering Branch, fell into two parts. First, part of the future commercial liquid high-level radioactive waste should be solidified and then permanently stored or buried in selected geological location. Second, the other part of the future commercial liquid high-level wastes should be directly disposed into such geological formations as salt structures and deep basins. The report rejected the location of future nuclear installations near geological sites suited for waste disposal and argued that "power reactor location, with present power transmission practices and economics, is predicated to a substantial extent on location of load" (USAEC 1957b, 18).[20] The location of chemical processing plants would primarily depend on the geographical concentration of reactors. Thus, within its task to aid and support the growth of the nuclear power industry, the Division of Reactor Development recommended site-selection criteria that were totally different to those proposed by the Hess Committee, although it adopted most of the other recommendations of both NAS committees.

The Hess Committee, already unhappy with the position of the Division of Reactor Development on site-selection criteria also became concerned about the geological aspects and consequences of other nuclear developments. Alerted by newspaper articles (and not informed by the AEC), the Hess Committee became concerned about project "Plowshare." This proposed that nuclear devices be detonated underground for peaceful purposes—for instance, creating underground storage space and excavating canals. In December 1958, Hess wrote to AEC chairman John McCone that this might lead to the release of large amounts of nuclear waste into the biosphere through ground water. The AEC commissioners were irritated by the letter and took the view that the Hess Committee had exceeded its area of competence. Their view was supported by the U.S. Geological Survey, which was actively involved in underground nuclear detonations and waste-disposal practices at AEC laboratories.

The 1959 Waste Hearings: Toward Stabilization

While the dispute between the Hess Committee and the AEC was developing, an institutional basis for a shared general problem definition was being laid at the AEC. The process started with the reports on radioactive waste of the Division of Reactor Development and its Sanitary Engineering Branch. Influenced by these reports and by interaction with Wolman, Gorman, and Lieberman, AEC Chairman Lewis L. Strauss suggested in September 1957 to AEC General Manager Kenneth Fields that the responsibility for waste disposal should be concentrated in one office or division.

Afraid that this concentration would restrict their waste-handling practices, production, and research programs, the other divisions, and notably the production sites under supervision of the Division of Production, resisted this proposal. They were supported by Fields, who argued that concentration would be impractical.[21] Fields did, however, create a Waste Disposal Working Group to discuss and coordinate the waste-disposal activities of the divisions. The working group met for the first time on January 31, 1958, under chairmanship of Lieberman.[22] After exploring the interests and responsibilities of the various divisions, the working group agreed to draft a basic AEC waste-disposal policy because of the "increasing need for a more formal integration on an interdivisional basis."[23] The group also decided to prepare and review plans and materials for a forthcoming hearing on nuclear waste disposal by the congressional Joint Committee on Atomic Energy (JCAE). In fact, the congressional hearing provided an external stimulus for the working group to compare, evaluate, and connect different views on and practices of radioactive waste disposal and to formulate a general AEC view that could be presented during the hearings. In addition, an informal steering committee was established to organize the hearings and to select and invite witnesses.[24] The preparation of the hearings by the Waste Disposal Working Group and the informal steering committee provided the Sanitary Engineering Branch with an additional opportunity to work out a shared view about radioactive waste management in the AEC.[25]

The Congressional Joint Committee on Atomic Energy had decided to hold hearings for several reasons. First, after several meetings with Gorman and Lieberman the Committee had become concerned about the implications of human health hazards and the cost of radioactive waste disposal for the development of the nuclear industry. Therefore it agreed with Gorman and Lieberman that the importance and the possibilities of additional research and development work, requiring larger budgets, should be identified and highlighted.[26] Second, the Joint Committee was afraid that, like fallout, nuclear waste disposal would be neglected. During the 1957 hearings on fallout several witnesses had argued that the dangers of radioactive waste were greater than those of fallout. Moreover, from 1957 onward, when the disposal of low-level radioactive waste in coastal waters became a major public, political and scientific issue in the coastal states, the public relations aspect of waste disposal was pushed on the AEC agenda. In the so-called dump debate the dangers of radioactivity from fallout, nuclear operations, and nuclear waste were compared and thus connected in a way that took the AEC by surprise (Mazuzan and Walker 1984; Divine 1978).

Hoping to calm such "emotional" and "nonscientific" fears, the Joint Committee presented the AEC with the opportunity of showing the outside world that reliable technical and scientific measures had been taken and that safe and long-term solutions for nuclear waste disposal were feasible. This opportunity was gladly accepted because, as Commissioner John Floberg wrote in June 1960, the commissioners were eager to allay anxieties, even when these were often "unreasonable, unfounded and scientifically unsound." According to Floberg, "occasional hysteria" and "sensational exaggerations in the press and over the air" could be best met "by taking extensive, even if not completely logical, measures to satisfy even unreasonable public doubt" (USAEC 1960b).

The Joint Committee hearings on industrial radioactive waste disposal were held in January and February 1959, and revealed some skepticism about waste management. In the opening statement, for instance, Wolman stressed the need for continuing governmental supervision because nuclear wastes were not a temporary problem, and he doubted that there was a lasting solution for high-level radioactive waste. Other critics stressed the technical complexity and the high cost of the treatment, storage, and disposal of high-level radioactive waste. However, most of the witnesses, representing AEC laboratories and private companies closely collaborating with these laboratories, testified, as they had done in the Wolman Committee, that despite many difficulties a technical, safe, and efficient solution for commercial high-level waste disposal was possible and would be available in the near future. These witnesses stressed that such solutions were being studied and developed in the AEC laboratories. All witnesses agreed that a solution for future commercial high-level radioactive waste was necessary for the unhindered growth of the nuclear power industry.

In conformity with the view of the SEB, Lieberman expressed his confidence that solutions for disposing future commercial high-level radioactive waste would be found before large-scale commercial nuclear power became a reality. He defined the most promising treatment and disposal solutions as interim tank storage, solidification, the separation of Sr90 and Cs137, followed by geological disposal of liquid and solidified waste. Because these solutions were feasible, Lieberman argued that a crash program was not necessary: We are not confronted with a situation where we have to take the position "we don't have time to do it right, so let's do it wrong, or let's pour more money into this thing so we can get it done tomorrow instead of the day after."[27]

Despite dissent by Wolman and a few others, the impression conveyed to Congress and the general public in these hearings was that there was no disagreement within the AEC on the definition, treatment, and disposal of high-level radioactive waste. This reflected the fact that all divisions, laboratories, and study groups in the AEC had moved toward a shared definition of the problem and the methods for its possible solution. Despite the fact that a long-term solution for the ultimate disposal of high-level radioactive waste had not been found, a research program had been started on treatment and disposal methods that were considered technically feasible and desirable in terms of safety and costs. This research was undertaken to guarantee the nonproblematic development of commercial nuclear power. During this research, existing treatment and disposal practices could be continued.

The AEC commissioners were satisfied with the outcome of the hearings, which coincided with their views about radioactive waste. In its Annual Report for 1959 the Commission proudly declared that

Waste problems have proved completely manageable in the operations of the commission.... There is no reason to believe that proliferation of wastes will become a limiting factor on future developments of atomic energy for peaceful purposes.... For the foreseeable future the commission will continue to store high-level wastes from its chemical separation plants in tanks.[28]

However, now that a shared view had been reached in the AEC, it had to be protected and strengthened so as not to impede the development of a suitable waste management technology.

Guarding the Stabilized Waste Definition

The stabilized view about radioactive waste problems and their solutions in the AEC was also noticeable in the Wolman Committee. In December 1959, Lieberman, the committee's rapporteur, distributed among the committee members "what is hoped will be the final draft of the summary report,"[29] in which he tried to reconcile the different problem definitions and solutions in the Wolman Committee. In this draft Lieberman stated that tank storage would not be replaced by geological disposal, but that both would coexist peacefully in any final disposal system. Furthermore, as he had argued during the 1959 hearings, there would be enough time to develop feasible disposal systems for the commercial liquid high-level waste before commercial nuclear power was introduced on a large scale. As the most satisfactory disposal approach, the draft report mentioned the conversion of high-level waste to a solid form, with subsequent storage of the solids in salt formations. The issue of disposal and dispersal practices at AEC laboratories also seemed to be solved:

"The cost of 'absolutely' processing or containing large volumes of low level wastes would be prohibitive and could present an unreasonable economic burden on the industry." (draft summary report, 17).

However, although the AEC proudly presented a reassuring view about radioactive waste management, the press highlighted the more disturbing features of AEC's nuclear operations. In the course of 1959, under the influence of the raging fallout debate, the dumping of low-level radioactive waste by private companies in U.S. coastal waters became highly controversial. Although the 1956 public report of the NAS Committee on Biological Effects of Radiation (BEAR) only had pointed at the relation between fallout and radioactive waste, the link between waste, fallout, and the genetic, biological, and pathological effects of atomic radiation became much more strongly connected in the public view during the debate on sea dumping.[30] The AEC was concerned by the scope and depth of the debate and the furious reaction of those living close to the dumping areas, because it believed the controversy might hamper the development of nuclear power. Accordingly, in 1960 the AEC commissioners decided not to issue new sea-disposal licenses.[31]

For much the same reason the continuing—but only internally expressed—criticism by the Hess Committee of waste-disposal practices at AEC laboratories was not taken lightly by the AEC. The Hess Committee had become frustrated because its recommendations did not reach the relevant laboratories. Many of its members were threatening to resign, and Lieberman was only able to avoid this with difficulty. In June 1960, Hess wrote a second letter to AEC chairman McCone criticizing the ground-disposal practices of the AEC laboratories and recommending both that waste disposal be carried out at suitable geological sites and that future AEC installations be built at such locations. Hess explicitly connected his criticism with his committee's safe disposal definition, saying that high-level radioactive waste should be completely isolated from the biosphere for a least 600 years.

Hess's letter received much attention within the AEC. The Division of Reactor Development wrote a report (USAEC 1960c) in which it made it clear that the disposal definition of the Hess Committee was unacceptable to the AEC.[32] Based upon this report, General Manager Alvin R. Luedecke replied to Hess by letter in January 1961; Hess was told that his committee had exceeded its charter by evaluating current practices, and that the AEC did not intend to adopt the expensive proposals concerning the relocation of existing fuel-element processing facilities because a practical and safe long-term waste disposal solution would be developed by the AEC laboratories. Furthermore, Luedecke accused the Hess Committee of exceeding its competence and authority because research on the biological and medical aspects of radioactivity was being performed by expert organizations such as the Federal Radiation Council, the National Committee on Radiation Protection, the International Committee on Radiation Protection, and the NAS Committee on Biological Effects of Radiation. Luedecke also stressed that research on geologic disposal was also being undertaken by DuPont, the University of Texas, Oak Ridge National Laboratory, the U.S. Geological Survey, the U.S. Bureau of Mines, and the American Association of Petroleum Geologists.[33]

To silence the Hess committee, the AEC commissioners proposed to the president of the National Academy of Sciences that the name of the committee be changed to emphasize its limited competence.[34] In June 1961, the Committee on Waste Disposal (the Hess Committee) was officially renamed by the National Academy as the Committee on Geological Aspects of Radioactive Waste Disposal. The Hess Committee was thus limited to advising on and evaluating engineering aspects of *geological research and development work* relevant to nuclear waste disposal.[35] Some of the Hess Committee members suddenly began to question the role they had been playing:

Lieberman insisted that they still wanted and needed the advisory committee, but I am inclined to think that they need it more as a political screen than for its nominal purpose of advising on waste disposal practices.[36]

Linn Hoover, Executive Secretary of the National Academy's Earth Sciences Division, summarized the outcome of the NAS-AEC decision by writing that he considered the Hess Committee "a lost cause."[37]

So, at the threshold of the 1960s the Hess Committee was silenced in order to preserve the generally shared view on radioactive waste and its existing and future management in the AEC. The success of this effort would further stabilize the social network around the defined problems and their solutions in the 1960s. The Sanitary Engineering Branch people were satisfied with the outcome; "The necessity for the persuasive and educational process was not as great after that."[38] In future they could devote most of their time, energy, and resources to research and development work.

Discussion

In this chapter I have shown that the radioactive waste problem we recognize today was socially shaped in the period 1945–60. Radioactive waste was defined in different ways: there were various radioactive wastes with entirely different problems, characteristics, applications, uses, and solutions. I would argue that the shared view of problems and solutions that was ultimately stabilized did not derive from the laws of nature, but instead was built on the contingent decisions, compromises, and negotiations between social actors. The radioactive waste that appeared to the American public in the early 1970s as a substantial problem was the result of a social and cognitive process in which heterogeneous elements—such as knowledge, assumptions, beliefs, persons, practices, policies, money, priorities, tasks, interests, power, and strategies—were brought together. Like the fluorescent lamp, the making of steel, clinical budgeting, and the construction of an airplane described elsewhere in this volume, radioactive waste was a social construction that was stabilized and given meaning in a complex frame of heterogeneous elements.

I have tried to show the heterogeneous character of technological development in large, complex organizations by focusing on the way in which relevant social actors tried to preserve, expand, and spread their way of seeing and treating radioactive waste inside the U.S. Atomic Energy Commission. In other words, the social shaping of a technology within an organizational structure was described by tracing the opportunities and constraints provided by the organizational structure to actors. Specifically, I have shown how the Sanitary Engineering Branch's efforts to increase attention to their definition of safe radioactive waste management were made possible by such resources as its organizational location, the second atomic energy act of 1954 aimed at the development of commercial nuclear power, and the mobilization and strategic use of external inputs provided by the fallout and sea-dumping controversies. However, the efforts of the SEB in the organizational setting of the AEC were also constrained, because it had to accommodate its strategies and goals to different AEC policies, priorities, power relations, and views on waste. This accommodation became clear as the task of the SEB gradually shifted to the development of a technology for future commercial waste, whereas initially it had tried to influence technologies for military wastes as well.

The organizational opportunities and constraints affected the strategies of the actors as they sought to advocate their views on radioactive waste in relevant fora, where all kinds of resources were mobilized, gained, and lost, and where different views on waste and its treatment were brought together. The creation of these fora constituted the essential first step in the strategy of the AEC actors to build a bureaucratic barrier between inside and outside, with the purpose of limiting the number of participants. In addition, these fora were also organizational arrangements that made it possible to link the inside and the outside (those would could speak and those who could not), by influencing those outside while denying them access to the inside. These fora thus successfully bureaucratized and technologized societal and political problems.

Fora created in this way represent both opportunities and constraints, because successful strategies aimed at the exclusion of the outside world do not completely determine the outcome of social interactions. This depends on the delicate interplay of other factors and conditions. In the radioactive waste case, for instance, the stabilization achieved in the fora was determined not only by insiders and their agendas but also by the ways in which the fora were shaped and whether their findings were transmitted to the outside.

The outcome of the processes described in this chapter was a stabilized view about radioactive waste, its problems, and solutions. Unlike nuclear power, radioactive waste was bureaucratized; but like nuclear power the stabilized view arrived at in the early 1960s was loudly proclaimed as a technological promise. This promise, which the outside world generally took to be appealing, can be seen as a successful method for freezing bureaucratic and political relations. The outbreak of the radioactive waste problem in the early 1970s clearly demonstrates that how long such a freeze can be maintained remains uncertain.

Notes

Work for this paper is part of a project funded by the Netherlands Organization for Scientific Research (NWO). The used empirical sources are primarily archival materials of the U.S. Atomic Energy Commission (AEC), the Congressional Joint Committee on Atomic Energy (JCAE), and the U.S. National Academy of Sciences (NAS), which are located at the U.S. National Archives, the Archives and the History Division of the U.S. Department of Energy (DOE), the Archives of the U.S. National Academy of Sciences, and at the U.S. Library of Congress.

I want to thank Brian Wynne, Wim Smit, and especially Arie Rip, Wiebe Bijker, and John Law for their many useful and stimulating comments on previous drafts of this paper.

1. The report, "Handling Radioactive Wastes in the Atomic Energy Program," (1949c) described the character and the origins of radioactive wastes in terms of

public health risks and costs. Both would be further reduced in actual and future research and development work, while the then current (military) waste handling practices were evaluated as safe, economic, and completely under control.

2. USAEC 1949a, Appendix B.

3. Interview with Joseph A. Lieberman, November 9, 1988, Bethesda, Maryland, and interview with Walter Belter, former colleague and successor of Lieberman in the SEB, November 11, 1988, Tensington, Maryland.

4. In other words: the plastic flow in salt formations was thought to predominate fracturing at a depth of 3,000 to 4,000 feet. Because of this the salt would effectively seal liquid or solid wastes and isolate them permanently from circulating water.

5. However, the report also stated "Some questions exist at this time in the minds of most members concerning the long-term safety of waste disposal as practiced on these [Hanford and Oak Ridge] sites if continued for the indefinite future" (USNAS 1957a: 3).

6. Interview with William Benson, former member of the Hess Committee, November 5, 1986, Washington, D.C.

7. The Reactor Safeguards Committee, in which Abel Wolman served as a member, made this recommendations in its report "Summary Report of the Reactor Safeguards Committee" (WASH-3), written in March 1950 but declassified in March 1957 (Balogh 1987: 188–189).

8. The other subcommittees were on: genetic effects of atomic radiation, meteorological aspects of the effects of atomic radiations, oceanography and fisheries, pathologic effects of atomic radiation, and effects of atomic radiation on agriculture and food supplies.

9. "... the defensiveness of operations people with respect to their current approaches to waste management, ... that the Wolman committee might well have said some things in the reports that would indicate or state specifically that there had to be some changes or improvements in what they were doing, those people did not like to hear that" (Lieberman interview, November 9, 1988).

10. They decided that: reactor accidents, accidents in handling and storing radioactive wastes, the effects of fallout, technical and administrative aspects of waste transportation, use and storage of radioisotopes, (nuclear) industrial growth, reprocessing, fuel and reactor types, the economics of waste handling and storage, and the actual status of waste disposal and dispersal, would be the subjects for study. During the conference, study groups were appointed that had to report on these subjects. The broad definition was greatly influenced by Wolman, who wanted to study the problem from different perspectives. (Transcripts of Wolman Committee meeting, February 23–24, 1956; records of the Committee on Disposal and Dispersal of radioactive Wastes, USNAS Archives, Washington, D.C.)

11. Report of study group g ("Proposals of panel on genetics on limitations of radiation exposure for general population and for occupational personnel"), by Parker (General Electric Co.—Hanford), Morgan (Health Physics Division—Oak Ridge National Laboratory), and Western (Division of Biology and Medicine), April 7, 1956. (Records of the Committee on Disposal and Dispersal of Radioactive Waste, USNAS Archives).

12. Members of the Wolman Committee study group f, "Issues requiring further study."

13. "... do not like interference in paragraph one that present disposal methods at national establishments is not hazardous. It may or may not be but this has to be determined." (Telefax from Hess to Wolman, May 24, 1956; Records of the Committee on Disposal and Dispersal of Radioactive Waste, USNAS Archives)

14. The AEC commissioners, and especially AEC Chairman Lewis L. Strauss, a staunch supporter of nuclear weapon testing (Pfau 1984), were satisfied with the reassuring conclusion that the biological effects of radiation from all peacetime activities were negligible and therefore would not hamper the development of nuclear technology or national security. The reports of the NAS Committee on Biological Effects of Atomic Radiation (BEAR) were incorporated in the 1957 fallout hearings. However, the press paid much attention to the intensified debate in the scientific community on the possible health effects of Strontium-90 that arose after the publication of the reports.

15. "I also find that the report Mr. Culler [Oak Ridge National Laboratory] and Dr. Mclain [Argonne National Laboratory] have prepared is being quoted as a report of the committee and to my knowledge this has never had a formal acceptance or designation." (Letter from committee member L. Silverman to S. D. Cornell, executive officer, USNAS, October 20, 1959. Records of the Committee on Disposal and Dispersal of Radioactive Wastes, USNAS Archives)

16. The criticism of the waste-handling practices at Oak Ridge National Laboratory in the second Wolman report reflected different competencies and definitions of waste treatment at ORNL: the chemical technology division (Culler) "versus" the health physics division (Morgan and Struxness), both represented in the Wolman Committee.

17. USAEC 1956a, "Disposal of radioactive waste," in USAEC 1956b, report 180-5.

18. USAEC 1956b, 3. Italics in the original. As most promising future alternatives to the expensive and unsafe (military) tank storage practices the authors mentioned: direct discharge in suitable deep geologic formations such as salt, shale, deep basins, and surface pits; fixation; and separation of Sr90 and Cs137. For the long-term future, the authors favored fixation and solidification of the commercial liquid high-level wastes, based on costs, transport, treatment, and control arguments. The economic use of radioisotopes was not excluded, but it was not stressed either. The authors emphasized the solubility of the anticipated problem by stating that if the requested money was given (1957: $1.5 million) to support "an aggressive research and development program we are confident that practical, safe ultimate disposal systems will be developed" (USAEC 1956b, 11).

19. USAEC, 1956c: WASH-408. In addition to Gorman and Lieberman, Wolman was cited as a co-author.

20. In their May 1956 WASH-408 report, Gorman, Lieberman, and Wolman had already stressed that in the AEC site selection criteria would be influenced only by the "economics" of industrial energy production and national security considerations.

21. The General Manager was the AEC's chief executive manager. Three of the first five General Managers were military men: Major General K. D. Nichols, Brigadier

General K. E. Fields, and Major General A. R. Luedecke. Fields had been Director of AEC's Division of Military Application from August 1951 until May 1955. (Titus 1986, 27; Hewlett and Duncan 1969).

22. Other group members were: E. F. Miller (Division of Production), R. J. Moore (Production), F. K. Pittmann (Office of Industrial Development), L. R. Rogers (Division of Licensing and Regulation), O. T. Roth (Division of Reactor Development), F. Western (Division of Biology and Medicine), E. van Blarcom (Division of Raw Materials), C. G. Marly (Office of Industrial Development).

23. USAEC 1958, 2. "It was the consensus that the WDWG can and should be a useful mechanism to promote and assure the integrated planning and assessment of operational and development programs in the field of waste management." (AEC 1958, 2) As subjects of integrated planning were mentioned: direct high-level waste disposal activities, commercially operated regional burial grounds for low-level wastes, assurance of adequate effluent control practices, sea disposal operations, and review of budget and manpower for waste disposal programs to assure adequate planning and support.

24. Members of this informal steering committee were: Abel Wolman, A. E. Gorman (SEB), J. A. Lieberman (SEB), W. Belter (SEB), H. M. Parker (General Electric—Hanford), E. G. Struxness (Health Physics Division, Waste Disposal Group-ORNL), H. Hanson (Director Robert A. Taft Sanitary Engineering Center, Cincinatti), C. W. Klassen (Chief Sanitary Engineer, Illinois State Department of Health, Division of Sanitary Engineering), P. Sporn (President American Gas and Electric Service Corporation), L. Hydeman/R. Löwenstein (Office of the General Council-AEC), D. Toll (JCAE staff), and J. T. Ramey (Executive Director JCAE).

25. The draft outline for the hearing had been made by the SEB, that is, "Lieberman and his staff" and was revised and expanded by the informal steering committee. (U.S. National Archives, Records of the Joint Committee on Atomic Energy (JCAE): Record Group 128, box 530, folder "Radiation, waste disposal hearings 1958."

26. Lieberman interview, November, 9, 1988; Belter interview, November 11, 1988.

27. JCAE, 1959. U.S. National Archives, Record Group 128, Records of the JCAE, Box 251, Volume: "Special Subcommittee on Radiation." During a meeting the Hess Committee reacted to this statement of Lieberman by declaring that "The committee is not suggesting a crash program; it is merely urging greater emphasis." (Minutes of meeting of the committee on waste disposal, May 14, 1960 at the Baker Hotel, Hutchinson, Kansas (USNAS Archives, Records of the Committee on (geological) disposal).

28. USAEC 1960a: 289, 299–300.

29. Letter from Lieberman to Wolman Committee members, December 18, 1959 (USNAS Archives, Records of the Committee on Disposal and Dispersal of Radioactive Wastes).

30. This connection was stimulated by the second fallout hearings in May 1959, during which most witnesses testified that renewed atmospheric testing would pose a serious threat to all mankind. This generally accepted conclusion totally differed from the outcomes of the 1957 fallout hearings and the conclusions formulated in the 1956 NAS report on the biological effects of radiation.

31. AEC, minutes Commission Meeting no. 1617, May 6, 1960; AEC, minutes Commission Meeting no. 1630, June 20, 1960. For an excellent account of the sea dump controversy and its relation to the fallout controversy, see Mazuzan and Walker 1984, 354–372.

32. "If it is interpreted to mean *zero* man-made radioactivity should be allowed in the environment, then any atomic activity would be virtually impossible" (USAEC, 1960c, 2).

33. USAEC 1960c: Appendix C; AEC, minutes Commission meeting 1675, November 23, 1960, letter from Luedecke to Hess, January 4, 1961 (USDOE Archives, Materials 12).

34. In his December 15, 1960 letter to Dr. Detlev Bronk, President of the National Academy of Sciences, Commissioner Robert E. Wilson stated that the commissoners constantly confused the Wolman Committee ("the top NAS-NRC committee") and the Hess Committee To end this confusion, Wilson suggested that "Dr. Hesses Committee should be renamed to indicate that it is a committee on the geological aspects of the disposal of highly radioactive waste. This, of course, is merely one suggestion as to possible methods of indicating how the scope of the committee could be clarified and confusion avoided" (USNAS Archives, Records of the Committee on (geologic) Waste Disposal).

35. The abolishment of the Hess Committee was not considered necessary: "In spite of philosophical convictions which appear to be held by certain Committee members, it is the opinion of the AEC waste development staff [the SEB] that the committee can continue to render valuable advisory service in the areas of ground disposal of radioactive wastes" (USAEC 1960d: Enclosure B, p. 10). In Lieberman's plain language: "You don't throw out the baby with the bath water" (Lieberman interview, November 9, 1988).

36. Letter of committee member King Hubbert to Linn Hoover, executive secretary of the Earth Sciences Division, NAS, June 30, 1961 (USNAS Archives, Records of the Committee on (geologic) Waste Disposal).

37. Letter from Linn Hoover to King Hubbert, July 13, 1961 (USNAS Archives, Records of the Committee on (geologic) Waste Disposal).

38. Lieberman interview, November 9, 1988.

Appendix 1: Divisional Competencies in the AEC

The AEC was a heterogeneous organization that encompassed many different actors and demands, competing issues and interests, different agendas, and a variety of decisions. Its organizational structure can be characterized as divided into two levels, a headquarters organization in Washington, D.C., and an operational organization consisting of nationally dispersed laboratories and research centers. AEC headquarters contained different divisions, each charged with specific tasks and responsibilities. The local laboratories, operated by universities and large private companies, were supervised by local AEC operational offices functioning as arms of AEC headquarters.

Although the major part of local operations and programs fell under the general responsibility of one headquarter division, other local R&D programs could be funded and supervised by other headquarter divisions. This rather opaque and complex organizational structure obscured the often contradictory and conflicting policies, competencies, and responsibilities.

Division	Tasks
Division of Production	Production of fissionable material, from procurement of raw materials to reactor fuels and weapon parts.
	Headquarter staff unit for supervision of all AEC construction and for development and procurement of radiation detection instruments.
	Administration of licensing system for source material, production facilities, and export of equipment.
	Supervision of its operation offices.
Division of Military Application	Development, production, and testing of atomic weapons.
Division of Reactor Development	Responsible for the development and testing of reactors, including equipment and processes for their safe and effective use. It included the Division of Engineering.
	Supervision of its operation offices.
Division of Engineering	Responsible for special engineering and related problems for the Division of Reactor Development, such as:
	formulation of R&D programs and policies.
	development of chemical and metallurgical processes.
	research, development, and testing of special reactor materials.
	the procurement, stockpiling, allocation, and control of reactor materials.
	return to nature or ultimate storage of reactor materials "on a controlled, hazard-free basis in conformance with established standards."
	Provision of staff advice and assistance to other divisions and offices of operations on sanitary engineering problems.
Division of Research	Responsible for physical research relating to Atomic Energy.
	Collaboration with Division of Biology and Medicine.
	Supervision of AEC isotopes program.

Division	Tasks
	Administration of programs of Cooperation with Office of Naval Research and National Research Council (NRC).
	Participation in planning and control of programs in the National Laboratories and other AEC installations.
Division of Biology and Medicine	Responsible for all biologic and medical research in the AEC.
	Supervision of measures in the operations of the atomic energy program to guard the health of AEC, contractor employees, and the public.
	Collaboration with the Division of Research.
	Supervision of the AEC isotopes program.
	Administration of programs of cooperation with the Office of Naval Research and with the National Research Council (NRC).
	Participation in planning and control of programs in the National Laboratories and other AEC installations.

Source: U.S. National Archives, Record Group 326: Records of the U.S. Atomic Energy Commission (USAEC)—Office Files of David E. Lilienthal, Subject files 1946–1950, boxes 12 and 16.

Appendix 2: Membership of the Wolman and Hess Committees

The Wolman Committee

Name	Scientific background	Institutional affiliation
Abel Wolman (Chairman)	Sanitary engineer	Johns Hopkins University
J. A. Lieberman (Rapporteur)	Sanitary engineer	AEC-SEB
A. E. Gorman	Sanitary engineer	AEC-SEB
C. W. Klassen	Sanitary engineer	Illinois State Department of Public Health, Division of Sanitary Engineering
C. P. Straub	Sanitary engineer	U.S. Public Health Service (ORNL-Health Physics Division)
L. Silverman	Industrial hygientist	School of Public Health, Harvard University
F. Western	Health physicist	AEC-Division of Biology and Medicine
K. Z. Morgan	Health physicist	Director, ORNL-Health Physics Division

Name	Scientific background	Institutional affiliation
H. M. Parker	Health physicist	Director, Radiological Sciences Department, General Electric Company, Hanford Work*
F. L. Culler	Chemical engineer	Director, ORNL-Chemical Technology Division
W. A. Patrick	Chemist	Johns Hopkins University
L. P. Hatch	Sanitary engineer	BNL-Nuclear Engineering Department
H. H. Hess	Geologist	Princeton University
C. V. Theis	Geologist	U.S. Geological Survey, Albuquerque, New Mexico
S. Krasik	Physicist	Westinghouse Atomic Power Department
S. Mclain	?	ANL Program Coordinator
S. T. Powell	?	Consulting Engineer
P. Sporn	?	President, American Gas and Electric Company, New York City
P. C. Aebersold	?	Director, AEC Isotopes Division, ORNL

Abbreviations: SEB = Sanitary Engineering Branch; ORNL = Oak Ridge National Laboratory; BNL = Brookhaven National Laboratory; ANL = Argonne National Laboratory
*from 1956 onward, manager, Hanford laboratories, General Electric Co., Richland.

The Hess Committee

Name	Institutional affiliation
Harry Hess (Chairman)	Head, Department of Geology, Princeton University
John N. Adkins	Director, Geophysics Branch, Office of Naval Research
W. E. Benson	Manidon Mining Corporation/Program Director, NAS-Earth Sciences Division
J. C. Frye	Director, Illinois State Geological Survey
W. B. Heroy	President, Geochemical Corporation, Dallas, Texas
M. King Hubbert	Staff Consulting Geologist, Shell Oil Company, Houston, Texas
R. J. Russel	Dean, Graduate School, Louisiana State University/Chairman, Division of Earth Sciences, NAS/NRC
C. V. Theis	Staff Scientist, U.S. Geological Survey, Albuquerque, New Mexico
W. R. Thurston (Secretary)	Assistant to the Director, U.S. Geological Survey

6

Artifacts and Frames of Meaning: Thomas A. Edison, His Managers, and the Cultural Construction of Motion Pictures

W. Bernard Carlson

In much of the historical and contemporary literature on technological innovation, inventors are characterized as problem-solvers. Such a characterization, I have often thought, is misleading in that it presumes that problems simply exist "out there," waiting for inventors to find and solve them. Just as stars do not exist in order that astronomers may name them, so there was no "telephone problem" in 1876 waiting for Alexander Graham Bell. Indeed, Bell's genius lay in not only devising a telephone but in constructing the problem of the electrical transmission of speech in the first place (Gorman and Carlson 1990). Clearly, one of the major lessons that scholars of technology can borrow from the sociology of scientific knowledge is an awareness of how scientists and inventors construct both nature and explanations of nature.

To apply this lesson to technology, it is useful to think about inventors not as problem-solvers but instead as bundles of solutions who construct problems suited to their unique skills and ideas. One can identify these bundles of solutions by looking for patterns both in the ways inventors work and in their creations (Hughes 1977, 1989). For instance, Thomas Edison often used many of the same electromechanical elements in his inventions, creating for himself a vocabulary of inventive building blocks (Carlson and Gorman 1990; Jenkins 1984). Likewise, Thomas P. Hughes (1971) has shown that Elmer Sperry was intrigued by the idea of feedback control and that he deliberately sought opportunities to apply this idea. Frequently, inventors are aware of their personal patterns and aptitudes and consciously shape opportunities or problems that allow them to capitalize on their strengths.

Yet inventors are not just bundles of technical solutions; they are also bundles of social solutions. Inventors succeed in a particular culture because they understand the values, institutional arrangements, and economic notions of that culture. Moreover, they are often willing and able not only to invent technological artifacts but also to modify the social and economic arrangements needed for that artifact to come into use. In inventing his steamboat in 1807, Robert Fulton solved two problems; first, using a steam engine to propel his vessel, and second, negotiating with the New York state legislature as to what speed was required for a successful steamboat. Fulton knew that his low-pressure boat would have a limited speed, and so with the help of his partner Philip Livingston he convinced the legislature to modify the terms under which they would award a monopoly for transportation on the Hudson River (Philip, 1985). Clearly, Fulton succeeded because he was able to join his artifact with new political arrangements.

Thus in the course of developing an invention, inventors combine technical and social solutions. They know that success comes from interweaving the social and technical in ways that make it impossible to unravel and separate the two. Put in more specific economic terms, they achieve this interweaving by securing patents, establishing a business for manufacturing and marketing, and attracting customers. Inventors seek profits and fame by linking their artifacts with social organizations for production and consumption.

Throughout this volume, scholars show how individuals link the social and technical in a variety of ways, especially from a sociological perspective. John Law and Michel Callon, for instance, show how the development of the TSR.2 was not only the design of a jet airplane but also the simultaneous establishment of a complex network of government agencies and private firms. In this essay, I wish to supplement the sociological perspective by examining invention from a cultural and cognitive viewpoint. How do broad cultural beliefs and social patterns create and reinforce cognitive patterns or ways of seeing the world? How do inventors design artifacts and establish business strategies in response to these ways of seeing the world? Historians and sociologists of technology have not fully investigated how inventors create and work within frames of meaning, to borrow a term from the work of Harry Collins and Trevor Pinch (1982).

I shall argue that inventors invent both artifacts and frames of meanings that guide how they manufacture and market their creations. Specifically, I shall examine the experiences of Thomas Edison in developing motion pictures. This is a interesting but ironic case; although Edison pioneered this communications technology and exploited it for thirty years, he and his company were eventually forced to abandon it in 1918. It would be easy to conclude that

Edison was simply "behind the times" or "out of touch with reality," yet this case challenges the historian to develop an explanation that interprets motion pictures from what might have been Edison's perspective or frame of meaning. What assumptions about the business world and customers did Edison use to construct motion picture technology? Once embedded in both hardware and strategy, how did these assumptions continue to inform the actions of his company?

In pursuing these questions, I am considering the interaction of cultural beliefs and class bias with business strategy and technological design. Because business and technological decisions are often seen as determined by narrow technical and economic considerations, let me state my conceptual position at the outset. My contention is that in any given culture there are many ways in which a technology may be successfully used. Although individuals often claim that they employ a technology in a way that optimizes the return on investment, at the time they make their decision there are often several alternatives with equivalent economic outcomes. To select from among these alternatives, individuals must make assumptions about who will use a technology and the meanings users might assign to it. These assumptions constitute a frame of meaning inventors and entrepreneurs use to guide their efforts at designing, manufacturing, and marketing their technological artifacts. Such frames thus directly link the inventor's unique artifact with larger social or cultural values.[1]

Let me emphasize that inventors and entrepreneurs must not only construct the hardware or artifact but simultaneously fashion frames of meanings. If they fail to do so, then they are often unable to sell their creation to investors and consumers. For instance, the successful development of electric lighting in the United States in the late nineteenth century depended on the linking of lighting systems with new assumptions about who would buy and use lighting equipment. In particular, inventors and entrepreneurs had to construct a new frame of meaning that focused on the creation of a new customer, the central station utility. As I have shown elsewhere, although Elihu Thomson was indeed a gifted inventor of lighting systems, his companies (Thomson-Houston and General Electric) succeeded because Charles A. Coffin linked Thomson's systems with the innovative strategy of central station utilities (Carlson 1991). Unable to sell lighting equipment in the same manner as steam engines or machine tools, Coffin instead helped local businessmen create a new form of company, the private electrical utility, and then sold equipment to this new customer. To pursue this new customer, Thomson added to his systems specific improvements suited to the needs of utilities. Thus, the rapid development of electric lighting in the United States cannot be understood solely in terms of technical developments; one must instead look at how new hardware was linked with the creation of a new business strategy, selling equipment to the newly formed central station utilities. For Thomson and Coffin, "the central station as customer" was a frame of meaning that they constructed as they designed and marketed their lighting systems.

Drawing on recent work in American cultural history, I will describe how Edison and his managers developed frames of meaning for motion picture technology that reflected cultural and social developments in two ways. First, Edison's own frame of meaning was shaped by the appearance of producer and consumer cultures. Scholars of American society have often viewed nineteenth-century America as a producer culture that celebrated the virtues of work, sacrifice, and perseverance. Its heroes were those men and women who tamed the frontier and created new technology and wealth. In contrast, twentieth-century America is marked by its consumer culture of leisure and indulgence. In this culture, the heroes are movie and sports stars, known primarily for their lifestyles. It is sometimes claimed that whereas the producer ethic was necessary to create the system of modern corporate capitalism, the consumer ethic is needed to provide ongoing demand for the products of the system. Furthermore, as corporate capitalism created a depersonalized and deskilled work environment, so average citizens responded by creating a compensatory culture of excitement and self-indulgence. I will argue that Edison developed his motion picture technology just as America was experiencing the transition from producer to consumer culture. Although Edison invented within the producer culture of the nineteenth century, his movie audiences and his competitors were participating in twentieth-century consumer culture.[2]

A second factor influencing Edison and his associates was class bias. After 1900, Edison delegated the motion picture business largely to his managers. As I will suggest, these men viewed the movies as a product for the middle class and shaped their business strategy accordingly Ultimately this strategy failed because in the teens the movies came to be a mass media, appealing to both the working and middle classes.[3]

I have deliberately chosen to use Collins and Pinch's term, frame of meaning, in this case study, but this term is closely related to several other concepts being developed in technology studies. Wiebe Bijker (1987; this volume) has employed the concept of technological frame to examine the "explicit theory, tacit knowledge, and general engineering practice, cultural values, prescribed testing procedures, devices, [and] material networks and systems" that social groups develop in relation to a specific artifact (introduction to this volume). Similarly, Bruno Latour (this volume) and Madeleine Akrich (this volume) use the concept of a script to denote the social behaviors that inventors and engineers design into a artifact. While either of these concepts could be used effectively to study Edison's motion picture inventions, neither exactly suits my purposes. On the one hand, Bijker's frame is too broad, encompassing too many important factors that shape how a social group assigns meanings to an artifact. I wish to focus on how cultural patterns and class bias informed the actions of Edison and his managers. On the other hand, the concept of a script is too narrow for this case, emphasizing social relations among users and between users and designers. While the script idea is a powerful analytical tool, it does not highlight how technologists draw on their larger culture to create an outlook or frame of meaning to guide their efforts. It is my sense that Collins and Pinch (1982) were trying to show how the culture of different scientific communities prepared different investigators to accept or reject evidence about parapsychology, thus making their concept appropriate for this chapter.[4]

Edison and the Culture of Production in the Nineteenth Century

For Americans, Edison is one of the great heroes of production. Along with Henry Ford, he is celebrated for having greatly contributed to the economic well-being of America through his inventions. Just as the story of George Washington chopping down the cherry tree is recounted to teach the importance of honesty, so stories about Edison are retold to emphasize the values of hard work, perseverance, and ingenuity (Robertson 1980). From an early age Americans learn how Edison stayed up night after night struggling to invent the incandescent lamp, and they are taught that Edison's favorite saying was, "Invention is 1 percent inspiration and 99 percent perspiration." Edison personalized the Protestant work ethic, revealing how one earns the respect of the community and contributes to the common good through hard work (Wachhorst 1981; Douglas 1987). Occasionally, Edison's active, productive efforts are contrasted with the self-indulgent and glamorous lifestyles of twentieth-century heroes of consumption.

It should not be surprising that Americans celebrate Edison as a hero of production, for Edison responded to the dominant values of his day and developed a production-oriented frame of meaning. Specifically, Edison's frame was aimed at business markets, avoided marketing to the general public, and looked to manufacturing for income. To see these characteristics, let us briefly review Edison's principal inventions.

Throughout his career, Edison preferred to develop inventions for use by business organizations, a preference he acquired early in his career with his telegraph inventions. During the 1870s, the completion of the national railroad and telegraph networks permitted some businessmen to manufacture and distribute goods nationwide. Anxious to tap this new national market, these businessmen welcomed communication innovations that increased the speed with which they received market news and the prices of stocks and commodities. As Alfred D. Chandler, Jr. (1977) has shown, a few businessmen used the telegraph to coordinate production and distribution functions within a single firm and thus created the first big business organizations. In response to these developments, Edison initially specialized in the creation of improved stock tickers and private-line telegraphs. Once established as a telegraph inventor, Edison improved the efficiency of the Western Union telegraph network. In particular, he introduced a quadruplex for sending four messages simultaneously over the same wire and a system of high-speed automatic telegraphy. Familiar with the needs of business offices, Edison then experimented with an early typewriter and introduced a duplicating system using an electric pen. Edison not only invented telegraph and business equipment, but he also established several factories in Newark, New Jersey for their manufacture.[5] Although he was becoming famous as an inventor, by the mid-1870s Edison was also, in his own words, "a bloated eastern manufacturer" (quoted in Josephson 1959, 85)

Working on these telegraph inventions convinced Edison of the value of inventing capital goods for a select business market. In general, Edison preferred to produce inventions that could be used by Western Union or other large firms. Frequently the managers of these firms knew what they wanted in communications technology —convenience and higher transmission speeds—and they were willing to pay for these improvements. In contrast, Edison learned through the experience of trying to sell an electric vote recorder to legislatures that it did not pay to develop inventions for which there were no preexisting social meanings; the vote recorder failed because legislators interpreted it as a threat to the practice of filibustering. More broadly, Edison perhaps sensed how difficult it could be to promote inventions to the general public. In the 1870s and 1880s, America was still a rural nation, consisting of thousands of small communities, each with its own values and mores (Wiebe 1967). To promote new technology in such a large and diverse market was an enormous effort, fraught with risk. Who could tell how individuals might want to use a new invention and what meanings they would bring to it? I think that Edison quite sensibly concluded that marketing a new technology to the general public was best done by businessmen who knew the local customs. As a result, Edison focused his efforts on producing machines for business markets and avoided marketing products to the masses. Wherever possible, Edison tried to make money by manufacturing his inventions and externalizing marketing and distribution.

Edison's preference for inventing capital goods and externalizing the marketing function can be clearly seen in the strategies he pursued with his two major inventions, the electric light and the phonograph. With the electric light, Edison designed a system that he expected to be used in offices, factories, and shops (Hughes 1983; Friedel and Israel 1986). Although he dreamed that electric lights would eventually be used in every home, Edison knew that this would only occur as the cost of lighting gradually decreased. Consequently, Edison focused his early efforts on selling the electric light to businessmen who had a need for artificial illumination and who could afford it. Edison did try his hand at building and promoting central stations in the mid-1880s, first with Pearl Street and then through the Thomas A. Edison Construction Department. However, he found that this work took him away from invention and required much negotiation with local groups to raise capital and to determine where and how stations should be operated (Hellrigel 1989). Edison again decided it was better to view electric lighting as a capital good to be sold to businessmen, who would either operate their own isolated lighting systems or establish utility companies. Although Edison helped the fledging utilities by making statements promoting electric lighting, his main business strategy was to make money by manufacturing equipment at plants in Harrison, New Jersey and Schenectady, New York.

In a similar manner, Edison applied his producer frame of meaning to the phonograph. During the 1870s, Edison produced a few tinfoil phonographs as novelties, but when the began full-scale manufacture a decade later, he intended that the phonograph be used as a business dictating machine. This decision informed the design of the phonograph of the late 1880s; rather than increasing volume, for example, Edison chose to enhance articulation so that typists would not miss words.[6] To distribute these machines, Edison and his associates set up an elaborate "state's rights" system in which agents purchased from Edison the right to sell phonographs in one or more states.

Edison might well have developed the phonograph for consumers, but he chose not to do so. Even though several of his competitors (most notably Emile Berliner) were selling phonographs for listening to prerecorded music, Edison regarded this as a wasteful application. According to Alfred O. Tate (1938), Edison's secretary and manager of his phonograph business, Edison only reluctantly permitted his phonograph to be used for "amusement" purposes after the business market failed. As Tate recalled (1938, 302), Edison took this position largely because of his producer values:

It is probable that this adaptation of the phonograph [to amusement purposes] was associated in his mind with the musical boxes so highly popular during the early Victorian era and broadly classified as "toys"....
His attitude indicates that he regarded the exploitation of this field as undignified and disharmonious with the more serious objectives of his ambition. He dedicated his life to the production of useful inventions. Devices designed for entertainment or amusement did not in his judgment fall within this classification. He did not desire that his fame, or any appreciable part of it, should rest upon a foundation of this nature.

At best, Tate was only able to convince Edison to develop a coin-operated phonograph for use in penny arcades. Here again, Edison insisted on selling the phonograph as a capital good to businessmen who would worry about promoting it to the general public.

Edison's experiences with telegraphy, electric lighting, and the phonograph all firmly established a producer frame of meaning in his mind, and it should not be surprising that nearly all of his later inventions were capital goods aimed at business markets. These included concentrated iron ore, Portland cement, primary and storage batteries, business dictating machines, and heavy chemicals. After 1900, Edison and his managers did promote the phonograph as a consumer good, but they encountered many of the same problems with this product as they did with motion pictures. Although the Edison organization was successful in bringing phonographs and music to a broad rural audience, it was unable to adapt to the rapidly changing tastes of urban and middle-class consumers.[7] In short, Edison's strength was in inventing machines that contributed to the "second industrial revolution" of the late nineteenth century.

To be sure, Edison's producer frame of meaning was not the only one that could be derived from the culture of late nineteenth-century America. Other inventors and entrepreneurs sensed that America was going through a transition from a producer to a consumer orientation and responded differently than Edison. Edison's friend and admirer Henry Ford was certainly steeped in the ethos of production, which drove him to revolutionize manufacturing. Yet Ford complemented his drive for mass production by addressing the problem of mass distribution, having his business manager James S. Couzens develop a network of franchised dealers (Rae 1965). Similarly, James B. Duke revolutionized the tobacco industry by introducing high-volume automatic cigarette-rolling machines, but to ensure adequate demand he had to create an organization capable of distributing and advertising his products worldwide (Chandler 1977). Clearly, Ford and Duke constructed frames of meaning that were strongly producer-oriented, but their frames also reflected the first signs of consumer culture. In contrast, Edison was much more like Andrew Carnegie, who concentrated on increasing efficency and lowering costs in the steel industry and did not concern himself with how steel was sold or used by consumers. Edison's producer frame of meaning was perhaps more narrow or rigid than that of other industrialists; nonetheless, it clearly reflected the dominant values of the period and was effective for many of his inventions

Edison and the Development of Motion Pictures, 1888–1900

Let us turn now to how Edison's producer frame of meaning informed his invention of the first motion pictures. Edison came to the idea of motion pictures by making an analogy with the phonograph. As he explained in an 1888 patent caveat, his motion picture machine or kinetoscope was to do "for the Eye what the phonograph does for the Ear, which is the recording and reproduction of things in motion".[8] Edison drew on this phonograph analogy in two ways. First, he used it to design his first kinetoscope as a machine that replaced the sound groove of the phonograph cylinder with a spiral of tiny photographs. Hoping to record and reproduce both sound and motion, Edison initially placed both the photographic and the acoustic cylinders on the single shaft of a machine similar to his phonograph. To view the moving images, Edison had the user peer through a microscope objective. This notion of a single viewer was similar to that employed by the existing phonograph, to which one listened through a set of individual eartubes. (Edison added the familiar loudspeaking horn to his phonograph in the 1890s). Even though Edison's assistant W. K. L. Dickson tested a crude projector in 1890, Edison insisted on developing the kinetoscope as a single-user device. Consequently, the first commercial kinetoscope was a peephole machine in which viewers watched the images through a small aperture (Carlson and Gorman 1990).

Second, the phonograph analogy informed Edison's marketing strategy for the kinetoscope. As with many new technologies, it proved easier to adopt this new invention to a preexisting marketing strategy than to pioneer a new scheme. Because phonographs were being sold for use in penny arcades, Edison permitted Tate and several other phonograph businessmen to establish similar kinetoscope parlors. Again, Edison established a "state's rights" distribution network in which agents purchased the rights to sell kinetoscopes in a territory, and these agents in turn sold machines to individual arcade owners (Allen 1982a). Under this strategy, kinetoscopes were manufactured in the Edison Phonograph Works, and Edison turned a profit by selling them outright to arcade owners. Initially, these machines cost about $50 to make, and Edison sold them for $100. During the 1890s, the Edison Manufacturing Company did a brisk business and sold more than 900 peephole machines.[9]

In the early 1890s, the public flocked to the kinetoscope arcades and marveled at seeing short films of boxers and vaudeville acts. These early films were shot at Edison's laboratory at West Orange under the supervision of Dickson and other staff members (Dickson 1933). Edison himself took little interest in these films and instead threw his energies into building a giant magnetic ore-processing plant in northern New Jersey; for him, this was a real "producer" invention (Carlson 1983). Edison saw little long-term potential in the kinetoscope, observing in 1894, "I am very doubtful if there is any commercial feature in it & fear that they will not earn their cost. These Zoetropic devices are of too sentimental a character to get the public to invest in."[10] Located in penny arcades alongside slot machines, phonographs, muscle-testing apparatuses, and fortune-telling machines, the kinetoscope seemed to Edison to be a frivolity (Peiss 1986). As a result this thinking, Edison decided to file only a few patent applications for the kinetoscope in the United States and none in foreign countries (Josephson 1959).

Although the public flocked to see the first kinetoscopes, they soon grew bored. In response, several kinetoscope exhibitors pressured

Edison to introduce a projecting machine and recapture the public's attention. In 1896, Edison relented and permitted his company to produce a projector based on a patent purchased from Thomas Armat.[11] During the remainder of the decade, the Edison Manufacturing Company sold over 800 projectors to small businessmen who exhibited films in vaudeville halls and makeshift theaters.[12] The Edison laboratory continued to make films on topics such as the beheading of Mary, Queen of Scots and the Battle of San Juan Hill in the Spanish-American War.[13] Significantly, Edison's associates do not seem to have worried as much about the artistic content of these films as they did about reducing production costs.[14]

Edison's Managers and the Motion Picture Industry, 1900–1918

Between 1903 and 1907 the American motion picture industry experienced several profound changes. All across the country, small businessmen began opening storefront theaters or nickelodeons where workers and immigrants could see a film for a nickel. Yet at the same time, American movie makers did not enjoy prosperity because the audiences in new nickelodeons preferred films made by British and French producers. In response, American filmmakers struggled to improve the media and as a result developed story films such as *The Great Train Robbery*. These two innovations—the nickelodeon and the story film—permitted entrepreneurs to market movies to a new broad audience, the urban working class. To do so, however, these entrepreneurs had to be sensitive to this audience's tastes and preferences (Sklar 1975; Allen 1982b; Rosenzweig 1983; Peiss 1986).

As motion pictures grew in popularity, the Edison organization was in a strong position. One of leading directors of the period, Edwin S. Porter, was their chief filmmaker (Jacobs 1939; Musser 1991). In 1905, to permit the production of films to keep up with demand, the Edison organization constructed a large studio in the Bronx in New York. By 1909, Edison had nine directors working at this studio and on location.[15] But most important in the minds of Edison and his associates was that, after several years of litigation, they won a series of favorable court decisions upholding the validity of Edison's patents on the kinetoscope. These legal victories were secured by Edison's attorney, Frank L. Dyer, who subsequently took over supervision of the motion picture business, first as Edison's chief counsel and then as president of Thomas A. Edison, Incorporated (TAE Inc.) (Ramsaye 1926; Cassady 1982).

From the outset, Dyer saw the patent victory as an opportunity for limiting the cut-throat competition in the motion picture industry. The success of the nickelodeons had stimulated the creation of thousands of theaters and about a dozen production companies, all competing to produce and exhibit the most exciting films. To bring order out of chaos, the Edison organization tried to use its patents to force all motion picture producers and exhibitors to take out licenses for their equipment. Dyer and other Edison managers insisted that it was not possible to construct either a motion picture camera or projector without infringing on Edison's patents.[16] In 1908, Dyer helped create the Motion Picture Patents Company (MPPC), through which the leading production companies pooled their patents and exerted some control over the industry by requiring all producers and exhibitors to have licenses.

Although the MPPC has been derided by some film historians as having harmed the evolution of the movies as a popular art form, Robert Anderson (1985) has argued that the MPPC had the important effects of eliminating destructive competition and permitting the rationalization of the industry. Through the MPPC, Dyer and other film industry leaders attempted to vertically integrate the industry to make it more stable, efficient, and profitable. Through a set of interlocking agreements, the MPPC controlled the supply of raw film, licensed the major film production companies and manufacturers of projection equipment, restricted the import of European films, coordinated film exchanges, and collected royalties from thousands of theaters. Anderson has suggested that by establishing uniform rental fees for all movies, the MPPC had the important effect of shifting competition among filmmakers from price to production quality.[17]

I would interpret the MPPC as an expression of the producer outlook. Through the MPPC, Dyer and the other leaders focused on the manufacture of films and less on developing movies as a form of mass entertainment. Their strategy of vertical integration was essentially the same as that being pursued by other giant firms intent on rationalizing steel production or automobile manufacture. Within the Edison organization, Dyer and other managers were successfully applying vertical integration to the manufacture of storage batteries and phonographs (Carlson 1988).

For the next few years, the MPPC figured prominently in the motion picture industry. At its height, MPPC's subsidiary, the General Film Company, controlled distribution of films to one half of the theaters in the United States. From 1911 to 1915, the Edison organization received one half of the MPPC's royalty and license fees or $1.9 million before expenses. Under these controlled market conditions, the Edison motion picture division enjoyed annual sales of over one million dollars.[19]

Having established a framework of vertical integration, Dyer and the Edison managers turned to shaping the content of their films. Their efforts reflected a middle-class bias; they viewed the movies as a product to be consumed by themselves or their social betters. Several tactics that reveal this bias. First, the Edison organization produced films that emphasized middle-class values and mores; the company was known for its wholesome comedies, biblical stories, and patriotic historical dramas. Typical Edison films in 1909 included a Thanksgiving Day release that contrasted the sacrifices of a pilgrim family with the problems encountered by a modern middle-class family; while the pilgrim family battles bears and Indians, the modern family "has adventures with swift-moving automobiles and the other current perils of a crowded street, arriving at their destination in a greater wreck than the ancient family." A second Edison film, *Annual Celebration of the Schoolchildren of Newark N.J.*, depicted "thousands" of schoolchildren at play in a beautiful Newark park "while teachers put sections of the scholars through graceful drills. All the children are dressed in white."[19]

Also illustrative of this middle-class orientation was the 1914 Edison release, *Andy Falls in Love*. In this picture a boy becomes infatuated with a theatrical actress and alters his personal grooming habits in the hope of wooing her. Too poor to purchase flowers for her, Andy resourcefully arranges to weed a neighbor's garden in order to pick a bouquet. At the climax, he presents the bouquet to his beloved, only to be thwarted by the actress's husband and adult son. Clearly, within the scope of a single reel, this movie offered lessons about passion, the cult of celebrity, personal hygiene, being resourceful, and the importance of the family. Thus we see how the Edison organization produced films that expressed the views of the middle class; the company stood in marked contrast to other filmmakers who were making popular romances, with hints of sex and violence.[20]

Second, rather than cater to the urban working class, Edison and his managers became concerned that the middle class was not patronizing nickelodeons. In response, Dyer attempted to have movie theaters opened in upper-middle-class towns near the Edison laboratory, such as his hometown of Montclair. However, by 1910 the movies had come to be viewed as a working-class amusement, and it was no surprise that the elite of Montclair refused to permit movies in their town.[21]

Third, the Edison organization supported the efforts to censor movies. Beginning in the mid-1900s, middle-class reformers were appalled to find that children and young women were frequenting the nickelodeons. The reformers were concerned about both the theaters as a near occasion of sin and the emotional and violent content of the movies. In response they passed local ordinances controlling the theaters, and they established the first motion picture censorship committee (Peiss 1986; Rosenzweig 1983). Dyer and the other Edison managers supported the censors, confident that their films would be approved because they reflected the proper values and interests of the middle class. Moreover, they believed that censorship was not only virtuous but should also be profitable; a full-page advertisement in *Moving Picture World* in December 1907 featured a quote from Edison: "In my opinion, nothing is of greater importance to the Success of the motion picture interests than films of good moral tone.... Unless it [i.e., the motion picture industry] can secure the entire respect of the amusement loving public it will not endure."[22]

One might well wonder why the Edison managers chose to produce movies with middle-class values while other companies produced movies for the burgeoning urban working-class audience. Why did they not pursue the largest segment of the market? One possible answer is that Edison's associates were affected by what historian Donald Finlay Davis (1988) has termed "conspicuous production." In reviewing how socially established families in Detroit created the automobile industry, Davis argued that aspirations of upward mobility led early manufacturers to produce vehicles for the well-to-do. "Each automotive entrepreneur," Davis noted, "built cars appropriate to his social background and present station in life. As he moved upward in the social hierarchy, his product climbed correspondingly in the industry's price-class hierarchy" (p. 8). With the important exception of Henry Ford (who came from outside the Detroit social aristocracy), the first generation of automobile entre-preneurs consciously avoided making cars for mass consumption and eventually lost out to Ford, General Motors, and Chrysler. In a like manner, the Edison managers may have had similar social aspira-tions that led them to produce movies for their middle-class peers and their social betters. Edison's associates knew that their social standing in the wealthy New Jersey suburbs of Montclair and South

Orange would not advance if it became known that their movies catered to the vulgar tastes of the working class.

Along with middle-class values, Dyer and the other managers were also influenced by Edison's producer values. This is especially apparent in their refusal to develop a star system. The star system was pioneered in the phonograph industry, in which Edison's prime competitor, the Victor Company, sold records by promoting individual performers such as Enrico Caruso. Edison opposed this practice, seeing it as simply giving in to the whims of egotistical performers (Millard 1990, 220). Consequently, unlike other film producers, the Edison managers did not cultivate celebrities to attract moviegoers (Jacobs 1939; Balio 1985). As one reads the correspondence from the motion picture division, one senses that the Edison managers were much more accustomed to producing capital goods such as storage batteries and supervising relatively taciturn workers; they were puzzled and annoyed by the behavior and demands of the actors and actresses.[23] At one point an Edison actress, Viola Dana, achieved a high degree of popularity and was compared favorably with Mary Pickford. However, aside from realizing that Dana might be as talented as Pickford, the Edison managers seemed to have no idea of how to promote her as a star.[24]

Not only did a producer orientation interfere with promoting stars, but it affected how the Edison managers handled other aspects of the movie business. Accustomed to production-oriented activities such as patent law, manufacturing, and engineering, Dyer and other Edison executives may have found many of the mundane tasks related to motion pictures peculiar and even distasteful. For instance, Edison managers devoted much time to reviewing dozens of scripts and securing copyrights to them. They also had to scrutinize photographs of potential actresses to determine whether they had the sort of eyes that "take well in motion pictures." Once a film went into production, they worried about the cost of delays on the set and the loss of costumes. Finally, Dyer became particularly concerned that films were not being properly edited and that the story lines lacked continuity. In response he established a film committee, made up of the chief Edison executives, to review all films before release. Week after week this committee met and agonized over cinematography and subtitle punctuation. Ultimately, however, this committee came to be driven by economic considerations, which only ensured mediocrity. As Leonard McChesney, the head of the motion picture division, complained in 1915,

We sit in the Film Committee week after week and pass pictures we know will get us nothing but unfavorable comments and cancellations. We haven't the power to throw out the distinctly bad pictures, nor the courage, because poor as they are they represent a certain sum of money invested in negative production. Four times out of five I leave the meetings feeling that I have had pictures jammed down my throat.[25]

As McChesney's remarks suggest, the Edison organization's two-pronged strategy of vertical integration and the infusion of middle-class values into movies eventually faltered. As the MPPC and the General Film Company sought to control more theaters, they angered the owners of independent theaters and film exchanges and attracted the attention of the Justice Department. Antitrust proceedings were begun in 1912, and the government formally ordered the dissolution of the MPPC in 1917. By then, however, the MPPC had lost most of its licensees, its income had been frittered away in numerous infringement lawsuits, and it was essentially defunct (Cassady 1982; Anderson 1985).

In the marketplace, Edison films also failed. Whereas prior to 1910 movies had been patronized largely by the urban working class, in the teens movies began to appeal to a mass audience of both the working and middle classes, immigrant and native-born Americans, country folk and city dwellers, men and women. Unfortunately for TAE Inc., movies without stars and emphasizing middle-class mores appealed to only a limited segment of this audience. Instead, this new mass audience preferred to see famous actors and actresses in movies with glamour, romance, and excitement. In large measure, this change in movie audiences was part of the transition from a producer to a consumer culture. After a day spent in an impersonal office or factory, Americans increasingly flocked to amusement parks, department stores, and movie theaters; through these institutions they compensated for the changes in their lives (Peiss 1986; Rosenzweig 1983). Both theater owners and filmmakers sensed this trend toward pleasure and entertainment, and they responded with more elaborate movie palaces, feature films, and stars. Thus the audience, filmmakers, and theater owners together constructed movies as a form of passive entertainment. In a larger sense, they used this technology to help create a new consumer culture that stressed celebrity, pleasure, and leisure. In contrast the Edison films, steeped in their producer and middle-class values, failed to reach this new mass audience.

As income from the motion picture division declined, Edison and his associates responded in predictable ways. True to his producer frame of meaning, Edison decided that the industry needed new hardware. Although Edison had dallied with color photography and a disk kinetoscope around the turn of the century, he and his associates devoted much energy after 1912 to improving motion picture technology.[26] Recalling his original dream of having talking images, Edison worked from 1912 to 1914 on a kinetophone that combined a projector with a special loudspeaking phonograph placed behind the screen. In this system, the projectionist controlled the phonograph by means of strings that sped up or slowed down the phonograph (Schifrin 1983a and 1983b). The kinetophone system proved unsuccessful because it was dependent on the skill of the operator and because theater rats liked to chew through the control strings.[27] Edison also introduced a smaller projector for use in churches, schools, and homes, which he called the home projecting kinetoscope. This product was probably a sound idea, in that a growing number of church leaders and social reformers viewed motion pictures as a desirable alternative to drinking and crime (Rosenzweig 1983). Yet as this product proved to be expensive ($75–100), it could only be afforded by a limited number of groups and well-to-do individuals and hence did not help solve the larger problem of attracting a mass audience for Edison motion pictures.[28]

Along with these new machines, Edison proposed a new direction in programming: educational films. Arguing that "the eye affords the quickest route to the brain," he ordered the preparation of an extensive series of films illustrating the basic principles of science. "I want to present the sciences and their application to industry and the related problems of life," Edison wrote in 1913. "I want to make the youth of this country unafraid of big things by showing them how big things are accomplished. I want to inspire in them a desire to do big things by filling their thoughts with big things."[29] Edison's goal for these films—to impart the values of producer culture—stood in marked contrast to other movie companies that saw their product as entertainment. To produce these educational films, Edison converted a portion of one of his lab buildings to a special studio and set up a special division of TAE Inc. Although Edison received much publicity for this scheme, it nonetheless failed to compensate for the loss of the mass audience for entertaining movies. In like fashion, another Edison manager tried to make a series of films for the Boy Scouts, but this deal fell through.[30]

As Edison and his associates experimented with new hardware and programming, the Edison organization also neglected to assess the impact of the new, longer feature films that non-MPPC filmmakers began introducing in 1914. These new films were four reels or longer, and with stars and substantial sets, they often represented an investment on the order of a several hundred thousand dollars per release (Jacobs 1939; Anderson 1985). For the Edison organization, given their commitments to a range of businesses (phonographs, storage batteries, and Portland cement), an investment of $100,000 per film was not possible and probably seemed ridiculous. At the time the typical Edison film cost between $1,000 and $5,000 to produce. Like the other movie companies, the Edison organization could have gone to Wall Street to raise this capital, but money borrowed for moviemaking might well have been money that the Edison group needed to finance its other enterprises. Consequently, the Edison motion picture division continued to "grind out" one- and two-reelers. Again, the production-oriented outlook emphasizing quantity production won out over the consumer-oriented outlook of modifying the quality of the films.[31]

The failure of these new machines and programming ventures, along with the decision ordering the dissolution of the MPPC, spelled the end of the Edison movie division. In 1916 the division stopped manufacturing projectors, and in 1918, after several poor years, Edison ordered the Bronx studio closed.[32] Thus although the Edison organization had survived longer than any of the other pioneer movie companies, it failed to adapt to the new world of movies as mass entertainment.

Conclusion

This case study shows well how the invention process involves the creation and linking of technological artifacts and frames of meaning. Edison's failure in the motion picture field previously has been attributed to several individual "wrong" decisions: his failure to pursue projection at the outset, his failure to secure adequate patent coverage for the kinetoscope, and his indifference to film production (Josephson 1959). Although each of these decisions was significant, Edison's attitude toward this new technology should be interpreted as resulting from the frame of meaning he applied to this invention. Throughout his career Edison insisted on inventing capital goods for businessmen, and he avoided becoming involved in marketing them to the general public. Consequently, as consumer culture emerged in the early twentieth century and the motion picture field turned sour for his company, Edison responded not with a new marketing scheme aimed at a mass audience but with new hardware and educational films. Likewise, just as Edison was guided by his producer values, so his managers were influenced by their middle-class outlook. Rather than seeing movies as a product for a mass audience, they insisted on producing films which narrowly reflected their own tastes and values. For both Edison and his managers, these frames were not expressed overtly, but they clearly shaped the design and implementation of the technology.

Both Edison's producer values and the middle-class bias of his managers ran counter to the emerging consumer culture. Other movie entrepreneurs discovered that Americans welcomed movies as a form of passive entertainment, and they strived to provide movies filled with new sensations and passions. To promote their films, these entrepreneurs established the star system, and they were willing to take the risk of introducing the multi-reel feature. For Edison and his associates, accustomed to producing capital goods such as storage batteries and heavy chemicals, the tasks of picking scripts for a mass audience and dealing with actors must have seemed alien, and they never mastered them.

It is important to note how these cultural values came to be embedded in the technology of the motion picture. Rather than inventing a new frame of meaning specifically for this artifact, both Edison and his associates used preexisting frames based on their previous experiences. Edison simply assumed that his kinetoscope would function much like his phonograph and would be marketed and used in the same ways. Likewise, Edison's managers assumed that motion pictures would be enjoyed by people like themselves, and consequently they emphasized middle-class values in the films. With the kinetoscope Edison transferred his frame of meaning from one machine to another; rather than consciously shape new meanings to fit the new technology, Edison let them "creep" into his design. Similarly, Edison's managers let their own middle-class background implicitly inform their decisions. Although we have come to expect that new technologies are revolutionary, I suspect that at the level of individual innovators and managers the process of cultural construction is often one of "cultural creep." By this I would suggest that inventors and producers often create artifacts to fit into cultural spaces suggested by their existing frames of meaning. Often, an inventor's survival depends on fitting into the existing order, not on consciously overthrowing it. It is only after the invention is put onto the market that consumers and other entrepreneurs use it in new ways and alter its cultural meanings.

The story of Edison and motion pictures raises questions about how Edison is portrayed in the scholarly literature. Frequently Edison is seen as the praiseworthy "heterogeneous engineer" who had the genius to link technical, social, political, and economic factors in his inventions (Law 1987b, Hughes 1983). To be sure, Edison did think carefully about the many external factors influencing his electric lighting system in the early 1880s, and he was able to reshape his storage battery in the 1910s to reach new markets (Carlson 1988). However, we should not assume that this was always the case for Edison. As the story of motion pictures demonstrates, Edison did not always function as a heterogeneous engineer who handled both the cultural and technical aspects of his inventions; here he chose to focus on the hardware at the peril of not fully understanding the social meanings the audience brought to this new technology.

Turning from Edison to his managers, I would make two observations. First, several historians of consumer culture have suggested that this culture was largely shaped by the elites and middle-class reformers as a means of controlling the growing working and immigrant classes (Fox and Lears 1983). Although this may be the case in terms of the new therapeutic outlook and mass-circulation magazines, I am reluctant to apply such intentions of controlling the working class to the managers of TAE Inc. To be sure, these managers were white men from the middle and upper classes, but I do not think they were especially concerned with using the movies to control the lower orders. They were interested in making money from the movies, and had they been able to understand consumer culture, they probably would have altered their product accordingly. However, like Edison, their strength lay with producer-oriented activities such as manufacture, patent law, and business organization, and they simply did not appreciate the trends of consumer culture. Not knowing what to do with this new technology, they made the reasonable assumption that it would be consumed by people like themselves. Although further research into the records of TAE Inc. may provide evidence to support the "control" thesis, I prefer to see the Edison managers as short-term profit-seekers, not long-term reformers.

Second, not only does this case cast doubts on the "control" thesis, but it also suggests that we need to rethink the process by which consumer culture appeared. I would suggest that the modern consumer world was created by the dialectical interaction of the working class and the elite. On the one hand, consumer culture was a "bottom-up" development, created by working-class audiences and entrepreneurs in response to the rapid changes in the workplace, the city, and the family. Workers and immigrants chose the amusements and activities that would permit them to cope with change, and entrepreneurs from working-class backgrounds pioneered these new services on a local level. On the other hand, elite corporate managers contributed to this culture by creating the business organizations capable of producing and distributing the goods, services, and values of consumer culture to a mass audience. Similarly, reformers reinforced elements of the new culture; for instance, they viewed movies as preferable to drinking and crime. However, the development of this dialectical thesis is well beyond the scope of this chapter; nevertheless, I hope that other scholars will investigate this perspective.[33]

Frames of meaning, heterogeneous engineering, and the dialectical interplay of elites and workers aside, in the final analysis Edison can be seen as the Moses of American consumer culture. It is true that he provided many of the basic technological artifacts of this new culture—the electric light, the telephone, and the phonograph—but he never understood the new culture that grew up around these devices. Edison led Americans to the Canaan of consumption, but steeped in his nineteenth-century values of production, he was unable to enter that promised land.

Notes

This chapter was written with the support of a summer research grant from the University of Virginia and the Newcomen Fellowship of the Harvard Business School, and I am grateful for this support. I conducted the archival research while A. J. Millard and I had a contract with the National Park Service. I wish to thank Edward J. Pershey, Mary B. Bowling, and Eric Olsen for their assistance in working with the collections at the Edison National Historic Site. I am especially grateful to Susan Douglas for suggesting this topic initially and for her valuable critique of the final version. Finally, thanks to Jeanne Allen, Charles Dellheim, Michael E. Gorman, Timothy Lenoir, Janet Steele, Jeffrey L. Sturchio, and Olivier Zunz, all of whom read and commented on this chapter.

1. This notion of locating the efforts of the inventor or entrepreneur in their social matrix is inspired in part by the entrepreneurial school of history that developed in the United States in the 1950s. As one entrepreneurial historian, Robert K. Lamb (1952) wrote,

[The entrepreneur] becomes a reality only when he is studied as a member of his society. The social groupings or institutions of that social system wherein he operates ... have their own value systems and goals which organize that society. Entrepreneurs, like other decisionmakers, depend for their success on the measure of acceptance their values and goals ... command from that society. (p. 116)

2. On producer and consumer culture, consult Fox and Lears 1983, Williams 1982, May 1985, McCracken 1988, and Susman 1984. Daniel Boorstin (1973, 89–164) provides a narrative overview of consumer culture.

3. For another example of class bias informing technological design, see Noble 1984.

4. I also wish to clarify the relationship between a frame of meaning and a mental model, a concept Michael E. Gorman and I (Carlson and Gorman 1989, 1990) are developing. In our research on the cognitive processes used by inventors, Gorman and I are investigating how inventors develop a conceptualization or mental model of an invention that they manifest in mechanical representations or specific physical devices. To date, we have found that for Alexander Graham Bell and Edison in the 1870s, their mental models appear to be one or more dynamic working devices that they manipulate in their imaginations and in sketches. In some cases, such as the kinetoscope, a mental model may include assumptions about manufacturing and marketing (Carlson and Gorman 1990). However, because we have not yet fully worked out the connections between marketing assumptions and the dynamic device, I have chosen to use frame of meaning in this chapter to focus attention on how an inventor draws on cultural patterns to guide both his business and technological efforts.

5. On Edison's telegraph and business inventions, see Jenkins et al. 1989, Jenkins and Israel 1984, and Jehl 1937.

6. "Edison's New Phonograph," *Electrical World*, Vol. 11 (7 Jan. 1888), 5.

7. Edison did manufacture phonographs and records in the early twentieth century, but I would argue that he was a follower rather than a leader in this industry. Instead, the Victor Company, under the leadership of Eldridge Johnson, was the pioneer in establishing the phonograph as a major form of consumer entertainment. For a discussion of how the Edison organization struggled to keep up with Victor, see Millard 1990, 208–216.

8. Edison to Seeley, 8 Oct. 1888, Patent Caveat 110, Cat. 1433, Edison Archives, Edison National Historic Site (hereafter cited as ENHS).

9. On the early manufacture of kinetoscopes, see Agreement between Edison and James Eagan, 26 June 1893, 1893 Motion Picture file; shop order 744, "25 kineto-graphs by Wm. Heise Jan," and shop order 779, "Making 1 punch, 1 cutter, 1 printing mach for E[dison] Mfg (Kinetoscope Dept) Sept [1894]," Notebook N871124; and Shipment [of kinetoscopes, Sept.–Nov. 1894, 1894 Motion Picture, Kinetoscope file, ENHS. For further information about the manufacture and promotion of kinetoscopes in the 1890s consult Hendricks (1966). The total number of kinetoscopes produced is from James H. White, testimony in "Complainant's Record," *Thomas A. Edison* vs. *American Mutoscope Company and Benjamin F. Keith*, U.S. Circuit Court, New York Southern District, In Equity No. 6298, 174, in legal box 173, ENHS. Hereafter cited as *Edison* vs. *American Mutoscope and Keith*.

10. Edison to E. Muybridge, 21 Feb. 1894, 1894 Motion Picture file, ENHS.

11. Norman C. Raff, testimony in *Edison* vs. *American Mutoscope and Keith*, 186–189.

12. The number of projectors produced is from James H. White, testimony in *Edison* vs. *American Mutoscope and Keith*, 174. For a discussion of how these projectors were used, see Allen 1982a.

13. Sklar 1975, 21 and *Newark News*, 15 Nov. 1902, 1902 Motion Picture Film Subjects file, ENHS.

14. On film costs, see A. T. Moore, "Film Report, 28 Feb. 1906–28 Feb. 1907," 1904 Motion Picture Sales file. As late as 1907, filmmakers still had to explain to top management the difference between "productions" using scenery and actors and documentaries that only had the expense of a cameraman. See A. T. Moore to W. E. Gilmore, 7 May 1907, 1907 Motion Picture Film file, ENHS.

15. E. J. Berggren to F. L. Dyer, 11 July 1910, 1910 Motion Picture file, ENHS; Jacobs 1939, 58–59; Ramsaye 1926, 440–441; and Millard 1990, 187.

16. T. Armat to W. E. Gilmore, 11 April 1903, F. L. Dyer to W. E. Gilmore, 21 July 1904, and no author to George Eastman, 20 May 1908, ENHS.

17. Jeanne Allen also provides a thoughtful assessment of the MPPC in her "Afterword" in Kindem (1982), 68–75.

18. The royalty figure is from Anderson 1985, 149, and the sales figure is from pocket notebook PN190101, ENHS.

19. Descriptions of these 1909 films are from *New York Variety*, 4 Dec. 1909, 1909 Motion Picture General file, ENHS.

20. See "Andy Falls In Love," [Oct. 1914], Records of Thomas A. Edison, Inc., Motion Picture Division, Box 2, Fol. 11, ENHS. For an overview of Edison films in this period, consult Slide 1970, 8–21.

21. F. L. Dyer, Memorandum, 13 Jan. 1910, ENHS.

22. See L. W. McChesney to C. H. Wilson, 22 May 1917, Dyer Correspondence, ENHS. Quote is from Singer 1988, 66, fn. 64. I am grateful to Leonard DeGraff for bringing the Singer article to my attention.

23. L. W. McChesney to M. R. Hutchison, 31 Jan. 1917, Dyer Correspondence, ENHS.

24. A. K. Watson to Edison Film Co., 17 June 1915, and L. W. McChesney to H. G. Plimpton, 22 June 1915, Dyer Correspondence, ENHS.

25. On the various mundane tasks, see the following items in the records of Thomas A. Edison, Inc., Motion Picture Division: H. Plimpton to L. W. McChesney, 22 Sept. 1914, Box 2, Fol. 23; C. H. Wilson to L. W. McChesney, 31 May 1916, Box 2, Fol. 16; F. L. Dyer to Edison, 17 Sept. 1912, Box 2, Fol. 47; and reports of film committee meetings, June–Dec. 1909, Box 10. Quote is from L. W. McChesney to C. H. Wilson, 2 Jan. 1915, ENHS. Miller Reese Hutchison also complained about the poor quality of Edison releases shown at committee meetings, characterizing them as "bum lousy pictures." See his diary entry for 2 Jan. 1914, ENHS.

26. On Edison's research into color photography, see J. H. White to A. Werner, 15 June 1900, 1900 Motion Picture File, and Edison to W. Heal, 9 May 1916, Letterbook LB160503, 52. On the disk kinetoscope, see Edison, entry for 18 March 1902, Notebook N020318. On additional experiments (such as stereoscopic projection) see M. R. Hutchison, "Status of Experiments" Notebook, N130227, ENHS.

27. On the difficulties of training operators, see "Kinetophone Difficult to Operate Says Inventor Thomas A. Edison," *Bayonne Review*, 23 May 1914, 1914 Motion Picture, Kinetophone file, ENHS.

28. On the home projecting kinetoscope, see Singer 1988 and L. W. McChesney, "The Home Kinetoscope," *The Edison Works Monthly*, Vol. 1, No. 2 (October 1912), 11–12 in Edison Papers, Box 48, Fol. 10, Archives and Library, The Edison Institute, Dearborn, Michigan. Hereafter cited as Edison Institute.

29. Quote is from Edison to Miss C. E. Mason, 31 March 1913, Edison Institute. Edison expected these films to be a major project; in 1911, the Edison organization began manufacturing 10,000 small projectors and made plans for several hundred films. See H. F. Miller to B. Singley, 21 Dec. 1911, Letterbook LB111204, 151, ENHS.

30. See record book of educational scientific film series, records of Thomas A. Edison Inc., Motion Picture Division, box 10 and Miller Reese Hutchison, Diary, entries for 28 August and 26 November 1913, ENHS.

31. On the Edison organization's decision not to produce multi-reel films, see L. W. McChesney, entry for 17 Dec. 1915, Pocket notebook PN151013, ENHS and L. W. McChesney to Scandaga Park Theatre, 28 Jan. 1914, Warshaw Collection of Business Americana, National Museum of American History, Washington, D.C. On film costs, cosult "Cost of Negative Film Subjects Completed by the Bronx Studio, November 1913," Incoming Correspondence file, 1913 Edison Company Correspondence, TAE Inc., Motion Picture Division, ENHS.

32. [W. Meadowcroft] to B. Singer, 26 July 1916, Letterbook LB160630, 338, and C. H. Wilson to C. Edison, 23 and 25 March 1918, 1918 Motion Picture File, ENHS.

33. One study that is sensitive to how the urban working class helped shape consumer culture is Kasson 1978.

III

What Next? Technology, Theory, and Method

The contributors to this volume share a commitment to the heterogeneity of social and technical relations. They are also committed to the view that sociotechnical change should be seen as contingent, and that it is, at least in part, a product of mixed strategies. But these commitments raise a series of questions. One of these is the question about where (or how) society ends and technology starts. How, if at all, can we disintinguish between the two? On this question there is less agreement.

We consider a number of possibilities more fully in the conclusion. Overall, however, it is possible to distinguish two approaches to the problem. One of these is what we might call the *interactive view*—a position characterized by three points. First, it is assumed that there is, indeed, a fairly stable and matter-of-fact division between the social and the technical. Second, it is assumed that the social shapes the technical. And third, it is reciprocally assumed that the technical is also capable of shaping the social. This view avoids the reductionisms of either social or technological determinism by arguing the case for interaction and exchange between the two. In this volume the authors who come closest to this view are, perhaps, Misa, de la Bruhèze, and Carlson.

However, there is a second and more radical approach—let us call it the *seamless web view*. This resists the notion that the division between the social and the technical is either stable or matter-of-fact. To say this is not, of course, to deny that it is possible to point at, and distinguish between, machines and those who operate them. Rather it is suggested that this distinction should be seen as an accomplishment, rather than something that can be taken for granted. Accordingly, it is argued that analysis should start with a seamless web of elements and look to see how that seamless web is broken up under different kinds of circumstances to create different *kinds* of objects. This seamless web approach is counterintuitive, but it is well represented in this volume. Notions like technological frame and actor-network, together with Bowker's study of Schlumberger, all assume that the social and the technical are constituted and distinguished in one movement—though this assumption is perhaps most fully developed in Bowker's paper.

But if sociotechnology is indeed a seamless web, then what kind of a vocabulary should we use in our analyses? The problem, as we indicated in the introduction, is that the language of common sense pushes us to talk of "technology" or "society"—as we have, for instance, above. It naturalizes the very distinctions that should be avoided by building them into the analysis instead of treating them as an object of study in their own right. This is the huge problem that Akrich and Latour seek to tackle. Indeed, their two papers, together with their joint "Summary," are best treated as a single piece. Their object is to press the merits of an evenhanded, relational semiotic language—one that allows them to escape the traps of common sense set by everyday speech.

Akrich uses this language to trace—in a way that resonates with several of the contributions in earlier sections—the manner in which the boundaries between the inside and the outside of sociotechnologies are delineated and thereby constitute what we commonsensically call technical artifacts and social actors. She exemplifies this with a series of studies of Third World electrification to illustrate the relationship between sociotechnical stabilization and the definition and distribution of attributions of agency and artifact. Thus, in the case of the electricity supply to Abidjan, she considers the various ways in which competences and moral attributions are distributed to different actors, human and nonhuman. The definition and formation of consumers, agents of the utility company, and electricity meters—each of these interactively plays a role in the process of stabilizing the network in question.

Latour has a similar concern with the distribution of competences between human and nonhuman actors. His first (deceptively simple) case is that of a door—its hinges, its operator, and its functions. He shows how tasks around the door may be delegated either to human or nonhuman actors—for instance, to a janitor or a mechanical "groom"—and explores the implications of these processes of delegation for others that interact with the door. He goes on to press the principle of generalized symmetry—the idea that agents and objects should be treated in the same terms—by exploring anthropomorphism in accounts of nonhuman actors and technomorphism in accounts of human actors. Finally, he considers the question of sociotechnical durability by distinguishing between the "programs" and the "antiprograms" that constitute and operate different versions of order in the semiotic seamless web.

If Akrich and Latour press a specific, symmetrical vocabulary for talking of and describing the seamless web of sociotechnology, then the paper by Pinch, Ashmore, and Mulkay leads us in another direction. Many recent studies have avoided a specific definition of technology—a matter that is considered head-on by these authors. First, drawing on a case study of health economics, they talk of what they call "social technologies"—procedures or methods of all kinds that have primarily to do with the engineering of social rather than technical relations.

Second, again drawing on their case study, they consider the rhetoric of technological formation—how it is that technologies come to be defined, tested, and evaluated. Here there are resonances with the work of Akrich and Latour, for the authors consider technology as a text and treat its rhetoric as a method by which the latter may gain persuasiveness and so stability. Their chapter considers the way in which two quite different rhetorics were deployed by protagonists of clinical budgeting in the U.K. National Health Service. Strong rhetoric, used primarily in discussions with economists, managers, and (market-oriented) politicians, defined and defended such budgeting as an effective tool for economic efficiency. The second "weak program" presented clinical budgeting as a user-friendly tool that would allow doctors to make decisions more effectively. Pinch, Ashmore, and Mulkay show that the protagonists of clinical budgeting switched strategically between the two repertoires—a process that traded on a distinction between "inside" and "outside" and by virtue of that fact tended to legitimate and so stabilize clinical budgeting.

Finally, the authors turn the spotlight on themselves. If health economists switch between rhetorical methods to legitimate their practice, then what implications does this have for the social analysis of technology? First, the authors note that their own accounts of clinical budgeting are not very different from those of the practitioners. For instance, in an earlier paper they characterized the two repertoires mentioned above as mutually incompatible. Thus *any* account of health economics—their own included—is just that, account that operates to stabilize or undermine the status of clinical budgeting as a social technology. But the point may be generalized. The process of juggling weak and strong vocabularies to keep them apart may be a widespread practice, not only in technology but also in the social technology of the sociology of technology. We are no different from those we claim to study!

7

The De-Scription of Technical Objects

Madeleine Akrich

Describing the Interaction between Technics and Humans

Although science and technology are often thought to go together, they are concerned with very different subject matters. Science is taken to go beyond the social world to a reality unfettered by human contingency. Perhaps as a result, the sociology of science has studied the ways in which the local and the heterogeneous are combined to create knowledge with the status of universal and timeless truth. By contrast, sociologists have found it difficult to come to terms with technical objects. Machines and devices are obviously composite, heterogeneous, and physically localized. Although they point to an end, a use for which they have been conceived, they also form part of a long chain of people, products, tools, machines, money, and so forth. Even study of the technical content of devices does not produce a focused picture because there is always a hazy context or background with fuzzy boundaries. Thus even the most mundane objects appear to be the product of a set of diverse forces. The strength of the materials used to build cars is a function of predictions about the stresses they will have to bear. These are in turn linked to the speed of the car, which is itself the product of a complex compromise between engine performance, legislation, law enforcement, and the values ascribed to different kinds of behavior. As a consequence, insurance experts, police, and passers-by can use the condition of the bodywork of a car to judge the extent to which it has been used in ways that conform to the norms it represents.

Technical objects thus simultaneously embody and measure a set of relations between heterogeneous elements. However, the process of describing everything about a car in such terms would be a mammoth task.[1] Furthermore, the end product might well be banal. The automobile is so much a part of the world in which we live that its sociography (a description of all the links making it up) would no doubt look like a collection of commonplaces. It would, in other words, look like a set of places where elements of the technical, the social, the economic, and so on were to be found together, and it would leave observers free to switch between one element or register and another as this suited them.[2]

I am arguing, therefore, that technical objects participate in building heterogeneous networks that bring together actants of all types and sizes, whether human or nonhuman.[3] But how can we describe the specific role they play within these networks? Because the answer has to do with the way in which they build, maintain, and stabilize a structure of links between diverse actants, we can adopt neither simple technological determinism nor social constructivism. Thus technological determinism pays no attention to what is brought together, and ultimately replaced, by the structural effects of a network. By contrast social contructivism denies the obduracy of objects and assumes that only people can have the status of actors. The problem is not one of deciding whether a technology should be seen as an instrument of progress or a new method for subjugating people. It is rather to find a way of studying the conditions and mechanisms under which the relations that define both our society and our knowledge of that society are susceptible to partial reconstruction.

To do this we have to move constantly between the technical and the social. We also have to move between the inside and the outside of technical objects. If we do this, two vital questions start to come into focus. The first has to do with the extent to which the composition of a technical object constrains actants in the way they relate both to the object and to one another. The second concerns the character of these actants and their links, the extent to which they are able to reshape the object, and the various ways in which the object may be used. Once considered in this way, the boundary between the inside and the outside of an object comes to be seen as a *consequence* of such interaction rather than something that determines it. The boundary is turned into a line of demarcation traced, within a geography of delegation,[4] between what is assumed by the technical object and the competences of other actants.

However, the description of these elementary mechanisms of adjustment poses two problems, one of method and the other of vocabulary. The difficulty with vocabulary is the need to avoid terms that *assume* a distinction between the technical and the social. Because the links that concern us are necessarily *both* technical and social, I develop and use a vocabulary drawn from semiotics that is intended to avoid this difficulty.[5] The methodological problem is that if we want to describe the elementary mechanisms of adjustment, we have to find circumstances in which the inside and the outside of objects are not well matched. We need to find disagreement, negotiation, and the potential for breakdown.

There are several areas—for instance, in technological innovation and technology transfer—where objects and their supposed functions, or the relationship between supply and demand, are poorly matched. In what follows I describe a number of cases of "technology transfer" to less-developed countries (LDCs) that are drawn from my own fieldwork. These range from the simple transplantation of a piece of technical apparatus widely used in industrial societies to the development of objects specifically intended for use in LDCs.[6] In each case I describe the elementary mechanisms of reciprocal adjustment between the technical object and its environment.

I start by considering the way in which technical objects define actants and the relationships between actants. I show that the ease with which the actants assumed in the design of the object are related to those that exist in practice is partly a function of decisions made by designers. The obduracy or plasticity of objects, something that is established in the confrontation with users, is a function of the distribution of competences assumed when an object is conceived and designed.

In the second part of the chapter I consider the way in which technical objects distribute causes. If most of the choices made by designers take the form of decisions about what should be delegated to whom or what, this means that technical objects contain and produce a specific geography of responsibilities, or more generally, of causes. To be sure this geography is open to question and may be resisted. Nevertheless, it suggests that new technologies may not only lead to new arrangements of people and things. They may, in addition, generate and "naturalize" new forms and orders of causality and, indeed, new forms of knowledge about the world. I will consider this process and illustrate the way in which technologies may generate both forms of knowledge and moral judgments.

Subjects and Objects in the Making

From Script to De-Scription

For some time sociologists of technology have argued that when technologists define the characteristics of their objects, they necessarily make hypotheses about the entities that make up the world into which the object is to be inserted.[7] Designers thus define actors with specific tastes, competences, motives, aspirations, political prejudices, and the rest, and they assume that morality, technology, science, and economy will evolve in particular ways. A large part of the work of innovators is that of *"inscribing"* this vision of (or prediction about) the world in the technical content of the new object. I will call the end product of this work a "script" or a "scenario."

The technical realization of the innovator's beliefs about the relationships between an object and its surrounding actors is thus an attempt to predetermine the settings that users are asked to imagine for a particular piece of technology and the pre-scriptions (notices, contracts, advice, etc.) that accompany it. To be sure, it may be that no actors will come forward to play the roles envisaged by the designer. Or users may define quite different roles of their own. If this happens, the objects remain a chimera, for it is in the confrontation between technical objects and their users that the latter are rendered real or unreal.

Thus, like a film script, technical objects define a framework of action together with the actors and the space in which they are supposed to act. Sigaut (1984) gives examples of tools whose form suggests a precise description (à la Sherlock Holmes) of their users. The two-handled Angolan hoe is made for women carrying children on their backs. The laborer's stake, with its single point, can only be driven in by two people, and thus presupposes a collective user. However, once one moves away from such simple examples, it becomes more difficult to uncover the links between technical choices, users' representations, and the actual uses of technologies. Thus the method of content analysis, as applied to texts, adopts an individual and psychological approach that has little or no relevance to our problem. Indeed, because it ignores the wide range of uses to which objects may be put, it comes close to technological determinism. It is obvious that it cannot possibly explain the wide variety of fates experienced by technological projects—fates that range from complete success to total failure.

One way of approaching the problem is to follow the negotiations between the innovator and potential users and to study the way in which the results of such negotiations are translated into technological form. Indeed, this method has been widely used in sociological and historical studies of technology. Thus, if we are interested in technical objects and not in chimerae, we cannot be satisfied methodologically with the designer's or user's point of view alone. Instead we have to go back and forth continually between the designer and the user, between the designer's projected user and the real user, between *the world inscribed in the object* and *the world described by its displacement*. For it is in this incessant variation that we obtain access to the crucial relationships: the user's reactions that give body to the designer's project, and the way in which the user's real environment is in part specified by the introduction of a new piece of equipment. The notion of *de-scription* proposed here has to be developed within this framework. It is the inventory and analysis of the mechanisms that allow the relation between a form and a meaning constituted by and constitutive of the technical object to come into being. These mechanisms of adjustment (or failure to adjust) between the user, as imagined by the designer, and the real user become particularly clear when they work by exclusion, whether or not this exclusion is deliberate.[8] The case of the photoelectric lighting kit is an example in which exclusion was explicitly sought by no one.

The Photoelectric Lighting Kit: Or How to Produce a Non-User

The photoelectirc lighting kit was born from the wish of a government agency to promote new energy sources. As part of its cooperative international activities, the agency wanted to work on and and meet the need for lighting—something that well-intentioned informants said was essential for all LDCs. At the same time it wanted to help the French photoelectric cell industry to create a market.

Caught up, as they were, in a specific network involving state support with industry, those involved in its design conceived of the kit as a function of the specific needs and constraints imposed on them by this network. At no point, for instance, did commercial considerations come into play. Accordingly, the shape of the lighting kit can be treated as a description of the way in which this network operated—a network characterized by the circulation of certain types of resources and the exclusion of other actors. The "narrative" patterns and scripts dreamed up by those who conceived the kits were quite specific, a function of their position. Study of the lighting kit (or any other technical object) makes it possible for us to create the "sociology" of the network defined by its circulation.

When I first heard the industrialists and designers talking about the lighting kit, it appeared to be a very simple array with three functional elements. There was a panel for producing electricity, a storage battery, and a lamp that consumed the electricity. However, once I arrived in Africa and started to study the ways in which such kits were actually used, the picture rapidly became more complicated. Those who were responsible for installing and maintaining kits were confronted with considerable difficulties. The first of these was that the wires linking the different components—the panel, the batteries, and the fluorescent tubes—were fixed in length and could not easily be altered because the connections were made with non-standard plugs. This meant that it was difficult to adapt the kits to fit rooms of different sizes. Replacing components with short lifetimes, such as lamps or batteries, represented a second set of difficulties. Neither appropriate fluorescent tubes, nor the watertight batteries chosen to ensure that maintenance problems would not limit the life of the system, were available in markets outside the capital. Local sources of supply were thus of no help to the user. As a result, despite the fact that it was a major element in his or her technical environment, the user lost control over the installation. Suddenly, what had previously been familiar started to become strange (the first question users asked was often "When do I have to add water to the batteries?"). A third factor also worked to prevent the user from appropriating the installation. This was the fact that the contractor who installed the kit forbade him or her to turn to a local electrician in case of breakdown. Instead, the contractor said that he would come to the area twice a year to repair faulty installations. The reason for this embargo on local repairs was the sensitivity of the photoelectric panel. This, as the instructions put it, "converts solar energy directly into electrical energy." However, the fact that this took the form of direct current with non-equivalent poles meant, at least in the view of the contractor, that it would be risky to call in a local electrician who would have experience of alternating but not of direct current. The danger was that if equipment was connected the wrong way, it might be damaged.

The discovery of these difficulties illustrates an important point of method. Before leaving Paris for Africa, the potential significance of nonstandard plugs, direct current, or waterproof batteries had not occurred to me. It was only in the confrontation between the real user and the projected user that the importance of such items as the plugs for the difference between the two came to light.[9] The materialization and implementation of this technical object, like others, was a long process in which both technical and social elements were simultaneously brought into being—a process that moved far beyond the frontiers of the laboratory or the workshop.

The fact that the importance of these characteristics only became evident in the interaction between designers and users was not the result of chance or negligence. Each decision actually taken made sense in terms of design criteria. Direct current is cheaper than alternating current because a transformer consumes a good part of the available power. Watertight batteries and nonstandard connections were chosen to prevent people from interfering with and so potentially damaging the kit. The length of the wiring had to be limited or it would reduce the performance of the equipment. These decisions were intended to ensure that the lighting kit would "work" under all circumstances—an important consideration in the negotiations between the industrialists and their clients. It should be recalled that it was not the latter who were the ultimate users of the kit, but rather the donating agency and the government to which the gift was to be made. Indeed, such was the concern to produce a foolproof kit that the designers decided not to have a separate switch in the circuit because this might become a point of illicit entry into the system. This meant that users often found it difficult to turn the light on or off because the only switch available was attached directly to the light and so was normally out of reach.

So it was that the technical object defined the actors with which it was to interact. The lighting kit (and behind it the designers) worked by a process of elimination. It would tolerate only a docile user and excluded other actors such as technicians or businesspeople who might normally have been expected to contribute to the creation of a technico-economic network. Had the users really been as docile as the designer intended, I would not have seen that the kit represented a large set of *technically delegated prescriptions* addressed by the innovator to the user.

If we are to describe technical objects, we need mediators to create the links between technical content and user. In the case of non-stabilized technologies these may be either the innovator or the user. The situation is quite different when we are confronted with stabilized technologies that have been "black boxed." Here the innovator is no longer present, and study of the ordinary user is not very useful because he or she has already taken on board the prescriptions implied in interaction with the machine. Under such circumstances some prescriptions may be found in user's manuals or in contracts. Alternatively, we may study disputes, look at what happens when devices go wrong, or follow the device as it moves into countries that are culturally or historically distant from its place of origin. In the next section I adopt the last of these methods to describe the use of generators in Senegal.

De-Scription in Technological Transfer: Reinventing and Reshaping Technical Objects in Use

In rural Senegal generators are widely used by "festive groups." An administration buys some small generators, which it distributes to youth groups in the villages. With the generators may come lights, a record player, or a loudspeaker. The youth groups use the generators or lend them to their members who pay for the cost of fuel and oil. Again, they may rent them out to other villagers who are also responsible for the cost of fuel and oil. The money that is made by the rental of generators is shared, with part going to the person who transports the generator and part going to the association. In this way a small collection of actors is involved with the generator—actors that can be seen as so many additions to the components that make up the generator.

The generator's metal trailer means that it is mobile, and so it plays an important part in this process. This is because the field of possible users and the relations between the different actors is defined by the movement of the generator. However, the fuel tank rivals the generator for the starring role because it draws a fundamental distinction between capital costs and operating costs. This distinction is inscribed from the outset in the social setup that brings the generator to the village: there is the administration, which underwrites the investment, and there is the group that actually manages and runs the generator. The technical device reduces negotiations between the two parties to a minimum because it directly suggests a pre-negotiated agreement. Obviously things could be arranged differently. This, however, would mean delegating a whole series of tasks to additional (legal, human, and technical) structures external to the generator and its trailer. It might even entail new systems of measurement—in which case it is not clear whether we would still be dealing with the same object.

The situation would be quite different if we were faced with a device whose costs were concentrated exclusively on the side of investment—as, for instance, with the photoelectric kits. What kind of relationship can there be between the buyer and the user under such circumstances? This was a question faced by those promoting the development of photoelectric cells in French Polynesia. Once these cells had been distributed, it was not always possible to insist that these two classes of costs should be distinguished. Not only did the technology itself fail to discriminate between them, but it offered no method of measurement that could be translated into appropriate socioeconomic terms. Thus no matter how it is used, a photoelectric panel generates current as a function of climate and latitude. The "standard" relationship between production and consumption (a reflection of the interdependence of two groups of actors) is replaced by an individual, direct, and indeed arbitary submission to natural forces.

The difference between this and the generator is obvious. In the case of the generator, the fuel tank can be used to measure the relationship between its use and the cost of that use—a relationship embodied in the motor as a whole. The creation of a particular kind of social link, that of renting out, is conditioned by the existence of this relationship, which delocalizes the generator by creating many groups of actors: investors/purchasers, owners/users, associate users, renters, and transporters. The existence of transporters makes the property even "purer," for they free it from servitude. Their payment marks the boundary of group solidarity, for the work of a single person cannot enrich the community. At the same time the generator builds a space and a social geography. Thus the teachers in one of the villages who needed lighting for their evening classes did not even consider renting a generator. The division between the world of the "market" and the "civic"[10] world may not have been brought into being in the village by the social differentiation entailed in electricity and its uses, but it was certainly modified by the latter.

The lighting kit put itself forward as a "hypothetical" object, whereas the generator was just another piece of equipment integrated into the various sectors of economic life. However, we should not overstate the difference between them. This is best seen in terms of differential resistance. It would would take much more effort to (re)dismantle the generator than it would the lighting kit. But in both cases we are dealing with the creation and extension of networks that simultaneously define both the social and the technical. Thus such items as nonstandard plugs and fuses become significant when the real users start to displace projected users. Again, the competence of the youth group, its relations with other elements of village life, the very definition of these elements—all of these are determined at the same time as, and by the same process, that defines the components that make up the generator. If we were to restrict our attention to the "function" fulfilled by this piece of equipment within the youth group, we might imagine that some other technical system (for instance, solar panels or connection to the national grid) would function in the same way. This, however, is not the case, for under such circumstances the relationship between the youth group and others in the village would be different and probably more fluid. In this sense, then, we can say that our relationships with the "real world" are mediated by technical objects.

Prescriptions as a Way of Enrolling Actors: Or How to Make Citizens

So far I have described technologies that appear to exercise relatively weak constraints over those who use them. If the generator and those who sponsor it nudge some who would otherwise be outside economic relations in the direction of involvement, then this effect is relatively small. In the case of the photoelectric lighting kit, the main danger is that no one will use it at all. However, technologies are not always like this. Sometimes their designers and builders use them to obtain access to certain actors, whom they push into specific roles. This is what happened in the case of the Ivory Coast and its electricity network. Here the physical extension of the network was an integral part of a vast effort to reorganize the country spatially, architecturally, and legally. The object was to create such new and "modern" entities as the individual citizen.

Winner (1980) has argued that certain technologies are inherently political—for instance, nondemocratic. If he is right about this, then the approach I have adopted here would lead to a form of technological determinism. However, the case of electrification in the Ivory Coast shows that even in those cases where there are marked political implications, it is first necessary to interest and persuade the actors to play the roles proposed for them.

Until recently village property in the Ivory Coast was collectively owned and under the control of elders, who allocated tracts of land to villagers as a function of their needs. This allocation was not permanent, and people might move to different areas. When the authorities started to think about electrification, they decided that this should be contingent on a more stable allocation of land, and in particular on a distinction between private and public property. Those developing the new electricity network (who also presented themselves as spokespersons for the general interest) assumed that the network would both contribute to this division and depend on it, as it would be installed on public land. In other words, the electricity network made it possible for the state to create its own space (the space of common interests) that could not be appropriated by anyone else. At the same time, it defined those with whom it would interact. Because only the individual would legally exist in this new system, former collective modes of village representation were thus systematically excluded.

To be sure, the creation of a system that allocated land permanently either to individuals or the state was a function of agreement in the village as a whole about the need for such stability. Through the new property system the electricity company was thus asking the villagers to make a *pre-inscription* witnessing their consent to a certain kind of future. Thus, individual villagers had to undertake certain formalities to secure title to fixed property. From the standpoint of the electricity company, legal ownership could be treated as a token for a range of agreements between different bodies about the future of the village. The new system of property was also the foundation for a series of projects by other utilities (the highway department, the water authority, the medical service, the education system). It meant that electrification could be integrated into various modernization programs, and it established economical procedures for consultation and political negotiation. Finally, the construction of the network itself would put the agreement of the village into practice and stablize it by making a durable inscription on the landscape.

But why should the villagers agree to enter into a game in which they would, or so it seems, lose a part of their independence? After all, by so doing they would place themselves under the influence of a central authority that would, by virtue of this very fact, increase its power. There are several answers to this question. The villagers wanted to have access to electricity. But there was the question of the way in which the company negotiated with the village. Indeed, to put it in this way is misleading. The company did not negotiate directly with the village. Rather, it negotiated with a spokesperson—invariably someone who had "succeeded" and moved from the village to the capital. Both this spokesperson, who negotiated with a range of central authorities on behalf of the village, and the villagers themselves knew that a series of indirect benefits would follow from agreement with the electricity company. After electrification the village could hope for better teachers, an improved health service, more financial support, and an increase in the number of development projects. In short, electrification was a method for avoiding direct and specific negotiations between the villagers and a series of external agencies. It was a package whose terms were fixed in advance. Those in the village had a choice. They could accept those terms or they could reject them, and overall the package was attractive.

In general an individual becomes a citizen only when he or she enters into a relationship with the state. In the Ivory Coast this was effected through the intermediary of cables, pylons, transformers, and meters. By contrast, in France individuals are inserted into such a wide range of networks that they have little chance of avoiding citizenship. From the registry office, via obligatory schooling to military service and the welfare state, the mesh of the state with its different superimposed networks draws ever tighter around them. In countries that have been created more recently, specific networks may come to the aid of a weak or non-existent state. The electricity network may create and maintain a relationship between an individual and a place. Thus in the Ivory Coast, where only a minority of salaried workers paid income tax, the electricity bill became the means by which local taxes were collected in recently built towns. Here, then, it was the electricity network that fostered a wider definition of the concept of citizenship.

From Causes to Accusations and Forms of Knowledge

In the examples above I have shown how technical objects define actors, the space in which they move, and ways in which they interact. Competences in the broadest sense of the term are distributed in the script of the technical object. Thus many of the choices made by designers can been seen as decisions about what should be *delegated* to a machine and what should be left to the initiative of human actors. In this way the designer expresses the scenario of the device in question—the script out of which the future history of the object will develop. But the designer not only fixes the distribution of actors, he or she also provides a "key" that can be used to interpret all subsequent events. Obviously, this key can be called into question—consumer organizations specialize in such skepticism. Nevertheless, although users add their own interpretations, so long as the circumstances in which the device is used do not diverge too radically from those predicted by the designer, it is likely that the script will become a major element for interpreting interaction between the object and its users.

Abobo-the-War and Marcory-No-Wire: Where Technology Meets Morality

In this section I focus on one particular process—moral delegation—and discuss devices installed by designers to control the moral behavior of their users. I describe the way in which such devices may measure behavior, place it in a hierarchy, control it, express the fact of submission, and distribute causal stories and sanctions.

As I have indicated, the introduction of the electricity network has established links between individuals in the Ivory Coast. The way in Which the individual/consumer relates to the network, and via the network to the electricity company, is codified and quantified by means of a basic technical tool, the electricity meter. This formulates the initial contract between the producer and the consumer. If one or the other fails to meet its obligations, the meter becomes invalid or inactive. Meters have a symmetrical effect on the producer/consumer relationship. The agreement of both is required if they are to tick over. Accordingly, the *set* of meters is a powerful instrument of control. Taken together, the set of meters measures the cohesion of the sociotechnical edifice materialized by the network. Consider the following story, which appeared in *The Kanian*, the electricity company newspaper, in its February–May 1985 issue:

OPERATION STRIKEFORCE AT "ABOBO-THE-WAR"
There is a flashing red light in the DR in Abobo, a lower class suburb of Abidjan, where there are 66,854 subscribers; the network's rate of return (the relationship between the energy put out by the producer and the energy billed to the clientele) has fallen from 0.93 to 0.87 in the space of one year!

Any reduction in the rate of return can be interpreted as an increase in the number of illicit connections, the work of corrupt employees, or a consequence of trafficking in meters. With both human and technical actors involved, the network measures illicit behavior and determines its character.

The definition of social space also extends to non-electrified areas. These are characterized in terms of their degree of deviance from the norm—that is, from electrification. Thus another suburb of Abidjan, Marcory, was split into two by the network. Each was given a name, and characterized in social terms:

Unlike residential Marcory, Marcory-No-Wire is a Marcory without electricity, without wires. It is well known that Abidjanis have a sense of humour. A suburb with no wires, imagine what kind of a spectacle that offers. For if electricity is a sign of progress, its absence suggests other absences: of hygiene in the streets, of buildings constructed to certain standards, of pharmacists, playgrounds, sportsgrounds and so on. When you add darkness at night to these absences, then the guardians of the peace would say you get a criminal haunt. (Toure 1985)

Even so, the dividing line between the permissible and the impermissible is negotiable. Thus in their strike-force operations, electricity company agents were told to replace so-called Russian meters that had proved defective without penalizing their owners, even though a simple tap on the meter would block it and allow unbilled electricity to be consumed. Unlike the agents, the "Russian" meters found it impossible to distinguish between licit and illicit behavior, between the actions of humans and nonhumans. Accordingly, although the contract between supplier and consumer remained in force, the meter failed in its prescribed role as the material inscription of that contract.

Each *individual* meter intervened as referee and manager of the relationship between supplier and consumer. Taken together, the *set* of meters operated as police in a collective organization, uncovering irregularities. Such irregularities appeared first as deviations in consumption curves that were neither localized nor sanctioned. They could, however, be quickly translated into "social" terms.

Some techniques move closer to "social control." They establish norms and punish those who transgress them. Thus the storage and regulation systems in photoelectric kits take the form of batteries and electronic components. The batteries store the electricity so that it can be given out, for example, for lighting when it is dark. However, the control system lies at the heart of a technical, economic, and social imbroglio. If the battery is allowed to run too low, its lifetime will be reduced. On the other hand, if it is overcharged, electricity may leak back into it and ruin the photoelectric cell. Users might, of course, be given meters with which they could plan their electricity consumption while avoiding both of these dangers. In fact this solution is never adopted because the designers do not believe that users will allow the technical requirements of the system to overrule their immediate wishes. Again, the designers could choose to increase the capacity of the system to cope with the likely demands of the users. This, however, is a costly option. Accordingly, the designers adopt the third option of installing a regulator that cuts off the current to the user when the charge on the battery gets too low, and isolates the photoelectric panel when it gets too high.[11] As a result, a particular mode of consumption is imposed: the user cannot be too greedy, yet neither can he or she hope to compensate for excess consumption by prolonged abstinence. The penalty for breaking the rules—rules that are both social and technical—is immediate and abrupt: the current is cut off and is not reconnected until the battery is adequately recharged.

This method of regulation is designed to "groom" the user. It offers a set of rewards and punishments that is intended to teach proper rules of conduct. However, a flaw in the system is that there is no easy way to measure the charge in the battery. Voltage is only a rough indication. What should be done about this? A general who is not sure of the loyalty of his troops has two options. He may choose to do nothing. Or, like the designers in this case, he may redouble his precautions and disciplinary measures. Accordingly, as I have mentioned, a particularly inflexible system with nonstandard plugs was adopted. Thus while the control device was telling the user not to get too big for his or her boots, the nonstandard plugs were imposing even more draconian limitations on conduct. No bypass of the control device was permissible!

Even so, in French Polynesia the control device proved to be a shaky ally for the designers, because the users felt that its sanctions were arbitrary. The result was that they denounced it and expressed their displeasure by telephoning the electrician every time the system treacherously cut off the current while they were quietly sitting watching television. The electrician, who quickly became tired of doing repairs in the evening, tricked the system by installing a fused circuit in parallel with the control device. When the control device shut off the current, users could bypass it with the fuse, and the electrician would only be called out the following morning. The fused circuit thus marked the submission of electricians to the wishes of their clients and allowed them to be present by proxy instead of being summoned in person by irate users.

The precarious and makeshift character of the fuse makes it plain that some kind of intervention was necessary, even if it only took place after the event. In this particular trial it was the electricians who pleaded guilty. In fitting the fuse, they recognized that the control device and their clients were *both* right and moderated the judgments of the former in favor of the latter.

"The Order of Things and Human Nature": The Stabilization and Naturalization of Scripts

I have described several cases in which technical objects preformed their relationships with actors and vested them with what could be called "moral" content. Because roles and responsibilities are allocated, accusations and trials tend to follow. In principle, no one and nothing is protected from such denunciation. In the case of the electricity network, the users were accused of failing to respect the contract with the meter. However, the electricity company also accused some of the meters of failing to represent that contract. In the case of the photoelectric kits, it was the electricians, and indirectly the manufacturers, who found themselves in the dock through the agency of the control device. Indeed, the story of the kits can be read as a long series of reciprocal accusations. The industrialists tended to argue that if it didn't work (technically), this was because it had been misused (socially). The users, or those who claimed to be their representatives, argued that if it didn't work socially, this was because it had been misconceived *technically*. Here, then, we see an almost perfect "reversible reaction" that reveals the lack of a relationship, through the kit, between designers and users. The users did not interest the manufacturers; they were only important to the extent that they made it possible to go to the ministry of overseas development and seek support for a product that did not yet have a market. And in this interaction the kit did not actually have to do anything. Rather it was the *users* who were treated as an instrument for building a relationship between the manufacturers and the government.

In the case of the electricity network, the situation was quite different. It is difficult to imagine a plausible argument for illegal connection to the network—one in which the electricity network would stand in the dock. This is because the network configured a whole range of relationships. I have already mentioned the meter and the way in which it was related to the allocation of property. But relationships were structured by the network in many other ways. For example, it also tended to stabilize living space. This was because, for reasons of security and as a guarantee of solvency, only "permanent" structures were connected to the grid. And of course, once the grid was in place, new commercial networks for distributing electrical equipment quickly sprang up. Thus once it was established, the network tended to promote both physical and social stability. A wide range of elements were brought together and given substance. A small fringe group of "deviants" could not possibly hope to find the strength needed to outweigh the many actors bound together by the grid. Accordingly, the electricity company could call upon the meters to act as unequivocal spokespeople at will. A double irreversibility had been established—a material irreversibility inscribed in space and practice, and a directional irreversibility where accusations and charges could no longer be reversed. Obviously the two were intimately linked.

In this section I have argued that technical objects not only define actors and the relationships between them, but to continue functioning must stabilize and channel these. They must establish systems of causality that draw on mechanisms for the abstraction and simplification of causal pathways. In the case discussed above, the replacement of the "Russian" meters was very much part of this process—a process designed to make diagnosis automatic. Farther along the same path lies artificial intelligence.[12]

Conclusion: Toward the Constitution of Knowledge

Once technical objects are stablized, they become instruments of knowledge.[13] Thus when an electricity company sets differential tariffs for high- and low-consuming domestic users, for workshops, and for industrial consumers, it finds ways of characterizing and identifying different social strata. If it also chooses categories used in other socioeconomico-political network, then the knowledge it produces can be "exported." "Data" can thus be drawn from the network and transmitted elsewhere, for instance, to economists concerned with the relationship between the cost of energy or GNP and consumption. However, the conversion of sociotechnical facts into facts pure and simple depends on the ability to turn technical objects into black boxes. In other words, as they become indispensable, objects also have to efface themselves. I will illustrate this with an example drawn from Burkino-Faso.

Burkino-Faso is a developing country with a tiny electricity network. Over the past few years it has been government policy to electrify urban centers. The first problem for the engineers and technicians was to judge potential demand and decide how large the network should be. Two different approaches were adopted. The economic studies unit asked potential subscribers what price they would be willing to pay for electricity. This approach assumed that there was a relationship between supply and demand, and that consumption would vary inversely with price. The technical unit adopted a very different method. It drew maps of the towns, marked off the built-up areas, and noted the characteristics of the houses (whether large or small, permanent or temporary, and so on). On the basis of this map they designed a network that would be legally, economically, and technically feasible—a network that would make use of public space and serve only permanent buildings and government facilities.

The results obtained by the two approaches were quite different. In particular, the geographical and legal approach of the technical unit suggested the need for a far larger network than the market-led approach of the economic studies unit. The latter had acted as if there were no need for technical mediation between price and consumption. They assumed, that is, that this relationship was a fact of nature that would be given concrete form by the electricity network. In a sense they were led astray by the naturalization effect, which occurs when technical systems are completely integrated into the social fabric. It is only when the script set out by the designer is acted out—whether in conformity with the intentions of the designer or not—that an integrated network of technical objects and (human and nonhuman) actors is stabilized. And it is only at this point that this network can be characterized by the circulation of a finite number of elements—objects, physical components, or monetary tokens. Disciplines such as economics and technology studies depend on the presence of a self-effacing apparatus that lies outside their domains. Economists extract one kind of information from technical objects, technologists another. They are able to do this because such objects function in stable situations. The introduction of a new device can thus be assimilated, for example by economists, into the price/consumption relationship. The economy is not cut off from technology; there is no radical disjunction.

This is why it makes sense to say that technical objects have political strength. They may change social relations, but they also stabilize, naturalize, depoliticize, and translate these into other media. After the event, the processes involved in building up technical objects are concealed. The causal links they established are naturalized. There was, or so it seems, never any possibility that it could have been otherwise.[14]

We are ourselves no more innocent in this respect than anyone else. For we are able to say that technical objects changed, stabilized, naturalized, or depoliticized social relations only with the benefit of hindsight. The burden of this essay is that technical objects and people are brought into being in a process of reciprocal definition in which objects are defined by subjects and subjects by objects. It is only after the event that causes are stabilized. And it is only after the event that we are able to say that objects do this, while human beings do that. It is in this sense, and only in this sense, that technical objects build our history for us and "impose" certain frameworks. And it is for this reason that an anthropology of technology is both possible and necessary.

Notes

I would like to thank Geoffrey Bowker, who translated this text, John Law, who carefully reviewed the entire text, and Bruno Latour, who helped me arrive at the more conceptualized form of the conclusions I drew from the various field studies discussed here.

1. Doubtless it could be satisfying to paint on a broad canvas, starting with nuts and bolts, pistons and cracks, cogs and fan belts, and moving on to voting systems, the strategies of large industrial groups, the definition of the family, and the physics of solids. In the case of such an inquiry we would no doubt find a mass of guides (people, texts, objects) ready to suggest ways in which we could extend our network. But such suggestions would be endless. On what grounds would the analyst stop— apart from the arbitrary one of lassitude? Quite apart from the indefinite amount of time such a study would take, there is also the question as to whether it would be interesting.

2. Here we are concerned with what might be called the consensual zone of the automobile, which is defined simultaneously by the major technical elements common to most vehicles and by their generally recognized uses. As is obvious, there are highly controversial zones around the margins, and it is around these points of friction that the battles leading to the establishment of supremacy of such and such a manufacturer or such and such a car are waged.

3. This term is used only as a convenient but imprecise shorthand. Depending on circumstances, the actor (a more general term to be prefered) may be a citizen, a member of a particular social class, a member of a profession, or even a finger or a body with a particular temperature as measured by a system of detection.

4. See Bruno Latour's text (this volume) for further discussion of delegation.

5. This vocabulary is further discussed in Latour's text in this volume and in the joint appendix to our papers.

6. I am aware that the reader may be frustrated by the way in which these examples are used. Within a short article it is not possible to give full details. But as they are intended to exemplify an argument, I hope that the reader will agree that the benefit of using them in this way outweighs the costs.

7. For a striking example of the interrelationship between the definition of technical parameters and the definition of a "world" for which the object is destined, see Callon's article on the electric vehicle in Bijker, Hughes, and Pinch 1987.

8. See, for example, Winner 1980 and Latour 1988a. Winner describes how the height of overpasses on the Long Island Parkway was chosen to prevent the passage of buses, the mode of transport most used by blacks, so that the use of leisure zones was effectively limited to whites. Latour, reinterpreting the example described by Daumas (1977), tells how, in exactly the same way, the radical Paris city council at the end of the nineteenth century decided to build metro tunnels too narrow for standard railway company trains. The objective, which succeeded for seventy years, was to prevent the private railway companies (supported by the right) from getting their hands on the Paris metro, whatever party happened to be in power. Multiple translations are necessary in order to arrive at such results. In Winner's case we need to move from the white/black to the car/bus distinction, and then on to the height of the overpasses. This is only possible because the black/white distinction is already *pre-inscribed* in unequal access to economic resources and, as a consequence, to expensive products such as cars. In Latour's case it is the width of the tunnels that allows the railway (and so the different companies and political parties) to be kept at arm's length from the metro.

9. In the French there is a play of words on *dessein* (design in the sense of plan) and *dessin* (design in the sense of drawing). The two have the same etymology.

10. I am drawing here on the distinction between "marchand" and "civique" discussed by Boltanski and Thevenot (1987).

11. Naturally, the different parts of the system are reconnected automatically once conditions change.

12. The question of "breakdown" is relevant to this issue and deserves further consideration. A "breakdown" relates closely to the definition I have offered of a technical object. This is because it can only be understood as a part of practice—that is, as the collapse of the relationship between a piece of apparatus and its use. A breakdown is thus a test of the solidity of the sociotechnical network materialized by a technical object. The rapidity with which the search for the causes of breakdown can be completed is a measure of this solidity.

13. Perhaps it would be better to say that the stablization of a technical object is inseparable from the constitution of a form of knowledge of greater or lesser significance. This hypothesis is powerfully supported by the case described by Misa (this volume): there an industry, a market, and the notion about what was to count as "steel" were all constructed simultaneously.

14. As is well known, Foucault (1975) has described the links between the technology of the penitentiary, power relations, and new forms of knowledge.

8

Where Are the Missing Masses? The Sociology of a Few Mundane Artifacts

Bruno Latour

To Robert Fox

Again, might not the glory of the machines consist in their being without this same boasted gift of language? "Silence," it has been said by one writer, "is a virtue which render us agreeable to our fellow-creatures."

Samuel Butler (*Erewhon*, chap. 23)

Early this morning, I was in a bad mood and decided to break a law and start my car without buckling my seat belt. My car usually does not want to start before I buckle the belt. It first flashes a red light "FASTEN YOUR SEAT BELT!", then an alarm sounds; it is so high pitched, so relentless, so repetitive, that I cannot stand it. After ten seconds I swear and put on the belt. This time, I stood the alarm for twenty seconds and then gave in. My mood had worsened quite a bit, but I was at peace with the law—at least with that law. I wished to break it, but I could not. Where is the morality? In me, a human driver, dominated by the mindless power of an artifact? Or in the artifact forcing me, a mindless human, to obey the law that I freely accepted when I get my driver's license? Of course, I could have put on my seat belt before the light flashed and the alarm sounded, incorporating in my own self the good behavior that every- one—the car, the law, the police—expected of me. Or else, some devious engineer could have linked the engine ignition to an electric sensor in the seat belt, so that I could not even have started the car before having put it on. Where would the morality be in those two extreme cases? In the electric currents flowing in the machine between the switch and the sensor? Or in the electric currents flowing down my spine in the automatism of my routinized behavior? In both cases the result would be the same from an outside observer— say a watchful policeman: this assembly of a driver and a car obeys

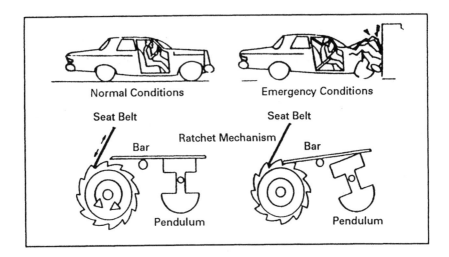

Figure 8.1
The designers of the seat belt take on themselves and then shift back to the belt contradictory programs: the belt should be lenient and firm, easy to put on and solidly fastened while ready to be unbuckled in a fraction of a second; it should be unobtrusive and strap in the whole body. The object does not reflect the social. It does more. It transcribes and displaces the contradictory interests of people and things.

the law in such a way that it is impossible for a car to be at the same time moving AND to have the driver without the belt on. *A law of the excluded middle* has been built, rendering logically inconceivable as well as morally unbearable a driver without a seat belt. Not quite. Because I feel so irritated to be forced to behave well that I instruct my garage mechanics to unlink the switch and the sensor. The excluded middle is back in! There is at least one car that is both on the move and without a seat belt on its driver—mine. This was without counting on the cleverness of engineers. They now invent a seat belt that politely makes way for me when I open the door and then straps me as politely but very tightly when I close the door. Now there is no escape. The only way not to have the seat belt on is to leave the door wide open, which is rather dangerous at high speed. Exit the excluded middle. The program of action[1] "IF a car is moving, THEN the driver has a seat belt" is enforced. It has become logically—no, it has become sociologically—impossible to drive without wearing the belt. I cannot be bad anymore. I, plus the car, plus the dozens of patented engineers, plus the police are making me be moral (figure 8.1).

According to some physicists, there is not enough mass in the universe to balance the accounts that cosmologists make of it. They are looking everywhere for the "missing mass" that could add up to the nice expected total. It is the same with sociologists. They are constantly looking, somewhat desperately, for social links sturdy enough to tie all of us together or for moral laws that would be inflexible enough to make us behave properly. When adding up social ties, all does not balance. Soft humans and weak moralities are all sociologists can get. The society they try to recompose with bodies and norms constantly crumbles. Something is missing, something that should be strongly social and highly moral. Where can they find it? Everywhere, but they too often refuse to see it in spite of much new work in the sociology of artifacts.[2]

I expect sociologists to be much more fortunate than cosmologists, because they will soon discover their missing mass. To balance our accounts of society, we simply have to turn our exclusive attention away from humans and look also at nonhumans. Here they are, the hidden and despised social masses who make up our morality. They knock at the door of sociology, requesting a place in the accounts of society as stubbornly as the human masses did in the nineteenth century. What our ancestors, the founders of sociology, did a century ago to house the human masses in the fabric of social theory, we should do now to find a place in a new social theory for the non-human masses that beg us for understanding.

Description of a Door

I will start my inquiry by following a little script written by anonymous hands.[3] On a freezing day in February, posted on the door of La Halle aux Cuirs at La Villette, in Paris, where Robert Fox's group was trying to convince the French to take up social history of science, could be seen a small handwritten notice: "The Groom Is On Strike, For God's Sake, Keep The Door Closed" ("groom" is Frenglish for an automated door-closer or butler). This fusion of labor relations, religion, advertisement, and technique in one insignificant fact is exactly the sort of thing I want to describe[4] in order to discover the missing masses of our society. As a technologist teaching in the School of Mines, an engineering institution, I want to challenge some of the assumptions sociologists often hold about the social context of machines.

Walls are a nice invention, but if there were no holes in them there would be no way to get in or out—they would be mausoleums or tombs. The problem is that if you make holes in the walls, anything and anyone can get in and out (cows, visitors, dust, rats, noise —La Halle aux Cuirs is ten meters from the Paris ring road—and, worst of all, cold—La Halle aux Cuirs is far to the north of Paris). So architects invented this hybrid: a wall hole, often called a *door*, which although common enough has always struck me as a miracle of technology. The cleverness of the invention hinges upon the hinge-pin: instead of driving a hole through walls with a sledgehammer or a pick, you simply gently push the door (I am supposing here that the lock has not been invented—this would overcomplicate the already highly complex story of La Villette's door); furthermore—and here is the real trick—once you have passed through the door, you do not have to find trowel and cement to rebuild the wall you have just destroyed: you simply push the door gently back (I ignore for now the added complication of the "pull" and "push" signs).

So, to size up the work done by hinges, you simply have to imagine that every time you want to get in or out of the building you have to do the same work as a prisoner trying to escape or as a gangster trying to rob a bank, plus the work of those who rebuild either the prison's or the bank's walls. If you do not want to imagine people destroying walls and rebuilding them every time they wish to leave or enter a building, then imagine the work that would have to be done to keep inside or outside all the things and people that, left to themselves, would go the wrong way.[5] As Maxwell never said, imagine his demon working *without* a door. Anything could escape from or penetrate into La Halle aux Cuirs, and soon there would be complete equilibrium between the depressing and noisy surrounding area and the inside of the building. Some technologists, including the present writer in *Material Resistance, A Textbook* (1984), have written that techniques are always involved when asymmetry or irreversibility are the goal; it might appear that doors are a striking counterexample because they maintain the wall hole in a reversible state; the allusion to Maxwell's demon clearly shows, however, that such is not the case; the reversible door is the only way to trap irreversibly inside La Halle aux Cuirs a differential accumulation of warm historians, knowledge, and also, alas, a lot of paperwork; the hinged door allows a selection of what gets in and what gets out so as to locally increase order, or information. If you let the drafts get inside (these renowned "courants d'air" so dangerous to French health), the paper drafts may never get outside to the publishers.

Now, draw two columns (if I am not allowed to give orders to the reader, then I offer it as a piece of strongly worded advice): in the right-hand column, list the work people would have to do if they had no door; in the left-hand column write down the gentle pushing (or pulling) they have to do to fulfill the same tasks. Compare the two columns: the enormous effort on the right is balanced by the small one on the left, and this is all thanks to hinges. I will define this transformation of a major effort into a minor one by the words *displacement* or *translation* or *delegation* or *shifting*;[6] I will say that we have delegated (or translated or displaced or shifted down) to the hinge the work of reversibly solving the wall-hole dilemma. Calling on Robert Fox, I do not have to do this work nor even think about it; it was delegated by the carpenter to a character, the hinge, which I will call a *nonhuman*. I simply enter La Halle aux Cuirs. As a more general descriptive rule, every time you want to know what a non-human does, simply imagine what other humans or other non-humans would have to do were this character not present. This imaginary substitution exactly sizes up the role, or function, of this little character.

Before going on, let me point out one of the side benefits of this table: in effect, we have drawn a scale where tiny efforts balance out mighty weights; the scale we drew reproduces the very leverage allowed by hinges. That the small be made stronger than the large is a very moral story indeed (think of David and Goliath); by the same token, it is also, since at least Archimedes' days, a very good definition of a lever and of power: what is the minimum you need to hold and deploy astutely to produce the maximum effect. Am I alluding to machines or to Syracuse's King? I don't know, and it does not matter, because the King and Archimedes fused the two "mini-maxes" into a single story told by Plutarch: the defense of Syracuse through levers and war machines.[7] I contend that this reversal of forces is what sociologists should look at in order to understand the social construction of techniques, and not a hypothetical "social context" that they are not equipped to grasp. This little point having been made, let me go on with the story (we will understand later why I do not really need your permission to go on and why, nevertheless, you are free not to go on, although only *relatively* so).

Delegation to Humans

There is a problem with doors. Visitors push them to get in or pull on them to get out (or vice versa), but then the door remains open. That is, instead of the door you have a gaping hole in the wall through which, for instance, cold rushes in and heat rushes out. Of course, you could imagine that people living in the building or visiting the Centre d'Histoire des Sciences et des Techniques would be a well-disciplined lot (after all, historians are meticulous people). They will learn to close the door behind them and retransform the momentary hole into a well-sealed wall. The problem is that discipline is not the main characteristic of La Villette's people; also you might have mere sociologists visiting the building, or even pedagogues from the nearby Centre de Formation. Are they all going to be so well trained? Closing doors would appear to be a simple enough piece of know-how once hinges have been invented, but, considering the amount of work, innovations, sign-posts, and recriminations that go on endlessly everywhere to keep them closed (at least in northern regions), it seems to be rather poorly disseminated.

This is where the age-old Mumfordian choice is offered to you: either to discipline the people or to substitute for the unreliable people another delegated human character whose only function is to open and close the door. This is called a groom or a porter (from the French word for door), or a gatekeeper, or a janitor, or a concierge, or a turnkey, or a jailer. The advantage is that you now have to discipline only one human and may safely leave the others to their erratic behavior. No matter who it is and where it comes from, the groom will always take care of the door. A nonhuman (the hinges) plus a human (the groom) have solved the wall-hole dilemma.

Solved? Not quite. First of all, if La Halle aux Cuirs pays for a porter, they will have no money left to buy coffee or books, or to invite eminent foreigners to give lectures. If they give the poor little boy other duties besides that of porter, then he will not be present most of the time and the door will stay open. Even if they had money to keep him there, we are now faced with a problem that two hundred years of capitalism has not completely solved: how to discipline a youngster to reliably fulfill a boring and underpaid duty? Although there is now only one human to be disciplined instead of hundreds, the weak point of the tactic can be seen: if this *one* lad is unreliable, then the whole chain breaks down; if he falls asleep on the job or goes walkabout, there will be no appeal: the door will stay open (remember that locking it is no solution because this would turn it into a wall, and then providing everyone with the right key is a difficult task that would not ensure that key holders will lock it back). Of course, the porter may be punished. But disciplining a groom—Foucault notwithstanding—is an enormous and costly task that only large hotels can tackle, and then for other reasons that have nothing to do with keeping the door properly closed.

If we compare the work of disciplining the groom with the work he substitutes for, according to the list defined above, we see that this delegated character has the opposite effect to that of the hinge: a simple task—forcing people to close the door—is now performed at an incredible cost; the minimum effect is obtained with maximum spending and discipline. We also notice, when drawing the two lists, an interesting difference: in the first relationship (hinges vis-à-vis the work of many people), you not only had a reversal of forces (the lever allows gentle manipulations to displace heavy weights) but also a modification of *time schedule*: once the hinges are in place, nothing more has to be done apart from maintenance (oiling them from time to time). In the second set of relations (groom's work versus many people's work), not only do you fail to reverse the forces but you also fail to modify the time schedule: nothing can be done to prevent the groom who has been reliable for two months from failing on the sixty-second day; at this point it is not maintenance work that has to be done but the *same* work as on the first day—apart from the few habits that you might have been able to *incorporate* into his body. Although they appear to be two similar delegations, the first one is concentrated at the time of installation, whereas the other is continuous; more exactly, the first one creates clear-cut distinctions between production, installation, and maintenance, whereas in the other the distinction between training and keeping in operation is either fuzzy or nil. The first one evokes the past perfect ("once hinges had been installed ..."), the second the present tense ("when the groom is at his post ..."). There is a built-in inertia in the first that is largely lacking in the second. The first one is Newtonian, the second Aristotelian (which is simply a way of repeating that the second is nonhuman and the other human). A profound temporal shift takes place when nonhumans are appealed to; time is *folded*.

Delegation to Nonhumans

It is at this point that you have a relatively new choice: either to discipline the people or to *substitute* for the unreliable humans a *delegated nonhuman character* whose only function is to open and close the door. This is called a door-closer or a groom ("groom" is a French trademark that is now part of the common language). The advantage is that you now have to discipline only one nonhuman and may safely leave the others (bellboys included) to their erratic behavior. No matter who they are and where they come from—polite or rude, quick or slow, friends or foes—the nonhuman groom will always take care of the door in any weather and at any time of the day. A nonhuman (hinges) plus another nonhuman (groom) have solved the wall-hole dilemma.

Solved? Well, not quite. Here comes the deskilling question so dear to social historians of technology: thousands of human grooms have been put on the dole by their nonhuman brethren. Have they been replaced? This depends on the kind of action that has been translated or delegated to them. In other words, when humans are displaced and deskilled, nonhumans have to be upgraded and re-skilled. This is not an easy task, as we shall now see.

We have all experienced having a door with a powerful spring mechanism slam in our faces. For sure, springs do the job of replacing grooms, but they play the role of a very rude, uneducated, and dumb porter who obviously prefers the wall version of the door to its hole version. They simply slam the door shut. The interesting thing with such impolite doors is this: if they slam shut so violently, it means that you, the visitor, have to be very quick in passing through and that you should not be at someone else's heels, otherwise your nose will get shorter and bloody. An unskilled nonhuman groom thus presupposes a skilled human user. It is always a trade-off. I will call, after Madeleine Akrich's paper (this volume), the behavior imposed back onto the human by nonhuman delegates *prescription*.[8] Prescription is the moral and ethical dimension of mechanisms. In spite of the constant weeping of moralists, no human is as relentlessly moral as a machine, especially if it is (she is, he is, they are) as "user friendly" as my Macintosh computer. We have been able to delegate to nonhumans not only force as we have known it for centuries but also values, duties, and ethics. It is because of this morality that we, humans, behave so ethically, no matter how weak and wicked we feel we are. The sum of morality does not only remain stable but increases enormously with the population of nonhumans. It is at this time, funnily enough, that moralists who focus on isolated socialized humans despair of us—us meaning of course humans and their retinue of nonhumans.

How can the prescriptions encoded in the mechanism be brought out in words? By replacing them by strings of sentences (often in the imperative) that are uttered (silently and continuously) by the mechanisms for the benefit of those who are mechanized: do this, do that, behave this way, don't go that way, you may do so, be allowed to go there. Such sentences look very much like a programming language. This substitution of words for silence can be made in the analyst's thought experiments, but also by instruction booklets, or explicitly, in any training session, through the voice of a demonstrator or instructor or teacher. The military are especially good at shouting them out through the mouthpiece of human instructors who delegate back to themselves the task of explaining, in the rifle's name, the characteristics of the rifle's ideal user. Another way of hearing what the machines silently did and said are the accidents. When the space shuttle exploded, thousands of pages of transcripts suddenly covered every detail of the silent machine, and hundreds of inspectors, members of congress, and engineers retrieved from NASA dozens of thousands of pages of drafts and orders. This description of a machine—whatever the means—retraces the steps made by the engineers to transform texts, drafts, and projects into things. The impression given to those who are obsessed by human behavior that there is a missing mass of morality is due to the fact that they do not follow this path that leads from text to things and from things to texts. They draw a strong distinction between these two worlds, whereas the job of engineers, instructors, project managers, and analysts is to continually cross this divide. Parts of a program of action may be delegated to a human, or to a nonhuman.

The results of such *distribution of competences*[9] between humans and nonhumans is that competent members of La Halle aux Cuirs will safely pass through the slamming door at a good distance from one another while visitors, unaware of the local cultural condition, will crowd through the door and get bloody noses. The nonhumans take over the selective attitudes of those who engineered them. To avoid this discrimination, inventors get back to their drawing board and try to imagine a nonhuman character that will not *prescribe* the same rare local cultural skills to its human users. A weak spring might appear to be a good solution. Such is not the case, because it would substitute for another type of very unskilled and undecided groom who is never sure about the door's (or his own) status: is it a hole or a wall? Am I a closer or an opener? If it is both at once, you can forget about the heat. In computer parlance, a door is an exclusive OR, not an AND gate.

I am a great fan of hinges, but I must confess that I admire hydraulic door closers much more, especially the old heavy copper-plated one that slowly closed the main door of our house in Aloxe-Corton. I am enchanted by the addition to the spring of a hydraulic piston, which easily draws up the energy of those who open the door, retains it, and then gives it back slowly with a subtle type of implacable firmness that one could expect from a well-trained butler. Especially clever is its way of extracting energy from each unwilling, unwitting passerby. My sociologist friends at the School of Mines call such a clever extraction an "obligatory passage point," which is a very fitting name for a door. No matter what you feel, think, or do, you have to leave a bit of your energy, literally, at the door. This is as clever as a toll booth.[10]

This does not quite solve all of the problems, though. To be sure, the hydraulic door closer does not bang the noses of those unaware of local conditions, so its prescriptions may be said to be less restrictive, but it still leaves aside segments of human populations: neither my little nephews nor my grandmother could get in unaided because our groom needed the force of an able-bodied person to accumulate enough energy to close the door later. To use Langdon Winner's classic motto (1980): Because of their prescriptions, these doors *discriminate* against very little and very old persons. Also, if there is no way to keep them open for good, they discriminate against furniture removers and in general everyone with packages, which usually means, in our late capitalist society, working- or lower-middle-class employees. (Who, even among those from higher strata, has not been cornered by an automated butler when they had their hands full of packages?)

There are solutions, though: the groom's delegation may be written off (usually by blocking its arm) or, more prosaically, its delegated action may be opposed by a foot (salesman are said to be expert at this). The foot may in turn be delegated to a carpet or anything that keeps the butler in check (although I am always amazed by the number of objects that *fail* this trial of force and I have very often seen the door I just wedged open politely closing when I turned my back to it).

Anthropomorphism

As a technologist, I could claim that provided you put aside the work of installing the groom and maintaining it, and agree to ignore the few sectors of the population that are discriminated against, the hydraulic groom does its job well, closing the door behind you, firmly and slowly. It shows in its humble way how three rows of delegated nonhuman actants[11] (hinges, springs, and hydraulic pistons) replace, 90 percent of the time, either an undisciplined bellboy who is never there when needed or, for the general public, the program instructions that have to do with remembering-to-close-the-door-when-it-is-cold .

The hinge plus the groom is the technologist's dream of efficient action, at least until the sad day when I saw the note posted on La Villette's door with which I started this meditation: "The groom is on strike." So not only have we been able to delegate the act of closing the door from the human to the nonhuman, we have also been able to delegate the human lack of discipline (and maybe the union that goes with it). On strike . . .[12] Fancy that! Nonhumans stopping work and claiming what? Pension payments? Time off? Landscaped offices? Yet it is no use being indignant, because it is very true that nonhumans are not so reliable that the irreversibility we would like to grant them is always complete. We did not want ever to have to think about this door again—apart from regularly scheduled routine maintenance (which is another way of saying that we did not have to bother about it)—and here we are, worrying again about how to keep the door closed and drafts outside.

What is interesting in this note is the humor of attributing a human characteristic to a failure that is usually considered "purely technical." This humor, however, is more profound than in the notice they could have posted: "The groom is not working." I constantly talk with my computer, who answers back; I am sure you swear at your old car; we are constantly granting mysterious faculties to gremlins inside every conceivable home appliance, not to mention cracks in the concrete belt of our nuclear plants. Yet, this behavior is considered by sociologists as a scandalous breach of natural barriers. When you write that a groom is "on strike," this is only seen as a "projection," as they say, of a human behavior onto a nonhuman, cold, technical object, one by nature impervious to any feeling. This is *anthropomorphism*, which for them is a sin akin to zoophily but much worse.

It is this sort of moralizing that is so irritating for technologists, because the automatic groom is already anthropomorphic through and through. It is well known that the French like etymology; well, here is another one: *anthropos* and *morphos* together mean either that which *has* human shape or that which *gives shape* to humans. The groom is indeed anthropomorphic, in three senses: first, it has been made by humans; second, it substitutes for the actions of people and is a delegate that permanently occupies the position of a human; and third, it shapes human action by prescribing back what sort of people should pass through the door. And yet some would forbid us to ascribe feelings to this thoroughly anthropomorphic creature, to delegate labor relations, to "project"—that is, to translate—*other* human properties to the groom. What of those many other innovations that have endowed much more sophisticated doors with the ability to see you arrive in advance (electronic eyes), to ask for your identity (electronic passes), or to slam shut in case of danger? But anyway, who are sociologists to decide the real and final shape (*morphos*) of humans (*anthropos*)? To trace with confidence the boundary between what is a "real" delegation and what is a "mere" projection? To sort out forever and without due inquiry the three different kinds of anthropomorphism I listed above? Are we not shaped by nonhuman grooms, although I admit only a very little bit? Are they not our brethren? Do they not deserve consideration? With your self-serving and self-righteous social studies of technology, you always plead against machines and for deskilled workers—are you aware of *your* discriminatory biases? You discriminate between the human and the inhuman. I do not hold this bias (this one at least) and see only actors—some human, some nonhuman, some skilled, some unskilled—that exchange their properties. So the note posted on the door is accurate; it gives with humor an exact rendering of the groom's behavior: it is not working, it is on strike (notice, that the word "strike" is a rationalization carried from the nonhuman repertoire to the human one, which proves again that the divide is untenable).

Built-in Users and Authors

The debates around anthropomorphism arise because we believe that there exist "humans" and "nonhumans," without realizing that this attribution of roles and action is also a *choice*.[13] The best way to understand this choice is to compare machines with texts, since the inscription of builders and users in a mechanism is very much the same as that of authors and readers in a story. In order to exemplify this point I have now to confess that I am *not* a technologist. I built in my article a made-up author, and I also invented possible readers whose reactions and beliefs I anticipated. Since the beginning I have many times used the "you" and even "you sociologists". I even asked you to draw up a table, and I also asked your permission to go on with the story. In doing so, I built up an inscribed reader to whom I prescribed qualities and behavior, as surely as a traffic light or a painting prepare a position for those looking at them. Did you *underwrite* or *subscribe* this definition of yourself? Or worse, is there any one at all to read this text and occupy the position prepared for the reader? This question is a source of constant difficulties for those who are unaware of the basics of semiotics or of technology. *Nothing in a given scene* can prevent the inscribed user or reader from behaving differently from what was expected (nothing, that is, until the next paragraph). The reader in the flesh may totally ignore my definition of him or her. The user of the traffic light may well cross on the red. Even visitors to La Halle aux Cuirs may never show up because it is too complicated to find the place, *in spite* of the fact that their behavior and trajectory have been perfectly anticipated by the groom. As for the computer user input, the cursor might flash forever without the user being there or knowing what to do. There might be an enormous gap between the prescribed user and the user-in-the-flesh, a difference as big as the one between the "I" of a novel and the novelist.[14] It is exactly this difference that upset the authors of the anonymous appeal on which I comment. On other occasions, however, the gap between the two may be nil: the prescribed user is so well anticipated, so carefully nested inside the scenes, so exactly dovetailed, that it does what is expected.[15]

The problem with scenes is that they are usually well prepared for anticipating users or readers who are at close quarters. For instance, the groom is quite good in its anticipation that people will push the door open and give it the energy to reclose it. It is very bad at doing anything to help people arrive there. After fifty centimeters, it is helpless and cannot act, for example, on the maps spread around La Villette to explain where La Halle aux Cuirs is (figure 8.2). Still, no scene is prepared without a preconceived idea of what sort of actors will come to occupy the prescribed positions.

This is why I said that although *you* were free not to go on with this paper, *you* were only "relatively" so. Why? Because I know that, because you bought this book, you are hard-working, serious, English-speaking technologists or readers committed to understanding new development in the social studies of machines. So my injunction to "read the paper, you sociologist" is not very risky (but I would have taken no chance with a French audience, especially with a paper written in English). This way of counting on earlier distribution of skills to help narrow the gap between built-in users or readers and users- or readers-in-the-flesh is like a *pre*-inscription.[16]

The fascinating thing in text as well as in artifact is that they have to thoroughly organize the relation between what is inscribed in them and what can/could/should be pre-inscribed in the users. Each setup is surrounded by various arenas interrupted by different types of walls. A text, for instance, is clearly *circumscribed*[17]—the dust cover, the title page, the hard back—but so is a computer—the plugs, the screen, the disk drive, the user's input. What is nicely

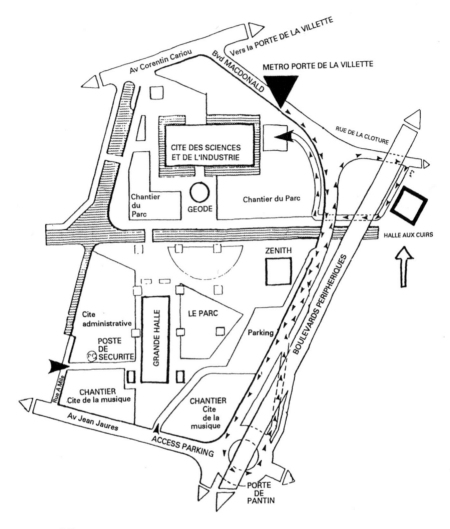

Figure 8.2
This is the written instruction sent through the mail by people from the Centre
d'Histoire des Sciences to endow their visitors with the competence of reading the
signs leading to their office, La Halle aux Cuirs. Of course it implies the basic
preinscribed competence: understanding French and knowing how to read a map,
and it has no influence on the other programs of action that lead people to want
to go to the Centre. It extends the mechanism of the door—its conscription—but
it is still limited in scope. Like users' manuals, it is one of those many inscriptions
that cover "the gap of execution" between people and settings.

called "interface" allows any setup to be connected to another through so many carefully designed entry points. Sophisticated mechanisms build up a whole gradient of concentric circles around themselves. For instance, in most modern photocopy machines there are troubles that even rather incompetent users may solve themselves like "ADD PAPER;" but then there are trickier ones that require a bit of explanation: "ADD TONER. SEE MANUAL, PAGE 30." This instruction might be backed up by homemade labels: "DON'T ADD THE TONER YOURSELF, CALL THE SECRETARY," which limit still further the number of people able to troubleshoot. But then other more serious crises are addressed by labels like "CALL THE TECHNICAL STAFF AT THIS NUMBER," while there are parts of the machine that are sealed off entirely with red labels such as "DO NOT OPEN—DANGER, HIGH VOLTAGE, HEAT" or "CALL THE POLICE." Each of these messages addresses a different audience, from the widest (everyone with the rather largely disseminated competence of using photocopying machines) to the narrowest (the rare bird able to troubleshoot and who, of course, is never there).[18] Circumscription only defines how a setup itself has built-in plugs and interfaces; as the name indicates, this tracing of circles, walls, and entry points inside the text or the machine does not prove that readers and users will obey. There is nothing sadder that an obsolete computer with all its nice interfaces, but no one on earth to plug them in.

Drawing a side conclusion in passing, we can call *sociologism* the claim that, given the competence, pre-inscription, and circumscription of human users and authors, you can read out the scripts non-human actors have to play; and *technologism* the symmetric claim that, given the competence and pre-inscription of nonhuman actors, you can easily read out and deduce the behavior prescribed to authors and users. From now on, these two absurdities will, I hope, disappear from the scene, because the actors at any point may be human or nonhuman, and the displacement (or translation, or transcription) makes impossible the easy reading out of one repertoire and into the next. The bizarre idea that society might be made up of human relations is a mirror image of the other no less bizarre idea that techniques might be made up of nonhuman relations. We deal with characters, delegates, representatives, lieutenants (from the French "lieu" plus "tenant," i.e., holding the place of, for, someone else)—some figurative, others nonfigurative; some human, others nonhuman; some competent, others incompetent. Do you want to cut through this rich diversity of delegates and artificially create two heaps of refuse, "society" on one side and "technology" on the other? That is your privilege, but I have a less bungled task in mind.

A scene, a text, an automatism can do a lot of things to their prescribed users at the range—close or far—that is defined by the circumscription, but most of the effect finally ascribed[19] to them depends on lines of other setups being aligned. For instance, the groom closes the door only if there are people reaching the Centre d'Histoire des Sciences; these people arrive in front of the door only if they have found maps (another delegate, with the built-in pre-scription I like most: "*you* are here" circled in red on the map) and only if there are roads leading under the Paris ring road to the Halle (which is a condition not always fullfilled); and of course people will start bothering about reading the maps, getting their feet muddy and pushing the door open only if they are convinced that the group is worth visiting (this is about the only condition in La Villette that is fulfilled). This gradient of aligned setups that endow actors with the pre-inscribed competences to find its users is very much like Waddington's "chreod":[20] people effortlessly flow through the door of La Halle aux Cuirs and the groom, hundreds of times a day, recloses the door—when it is not stuck. The result of such an alignment of setups[21] is to decrease the number of occasions in which words are used; most of the actions are silent, familiar, incorporated (in human or in nonhuman bodies)—making the analyst's job so much harder. Even the classic debates about freedom, determina-tion, predetermination, brute force, or efficient will—debates that are the twelfth-century version of seventeenth-century discussions on grace—will be slowly eroded. (Because *you* have reached this point, it means I was right in saying that you were not at all free to stop reading the paper: positioning myself cleverly along a chreod, and adding a few other tricks of my own, I led you *here* ... or did I? May be you skipped most of it, maybe you did not understand a word of it, o you, undisciplined readers.)

Figurative and Nonfigurative Characters

Most sociologists are violently upset by this crossing of the sacred barrier that separate human from nonhumans, because they confuse this divide with another one between *figurative* and *nonfigurative* actors. If I say that Hamlet is the figuration of "depression among the aristocratic class," I move from a personal figure to a less personal one—that is, class. If I say that Hamlet stands for doom and gloom,

I use less figurative entities, and if I claim that he represents western civilization, I use nonfigurative abstractions. Still, they all are equally actors, that is, entities that *do* things, either in Shakespeare's artful plays or in the commentators' more tedious tomes. The choice of granting actors figurativity or not is left entirely to the authors. It is exactly the same for techniques. Engineers are the authors of these subtle plots and scenarios of dozens of delegated and interlocking characters so few people know how to appreciate. The label "inhuman" applied to techniques simply overlooks translation mechanisms and the many choices that exist for figuring or defiguring, personifying or abstracting, embodying or disembodying actors. When we say that they are "mere automatisms," we project as much as when we say that they are "loving creatures;" the only difference is that the latter is an anthropomorphism and the former a technomorphism or phusimorphism.

For instance, a meat roaster in the Hôtel-Dieu de Beaune, the little groom called "le Petit Bertrand," is the delegated author of the movement (figure 8.3). This little man is as famous in Beaune as is the Mannekenpis in Brussels. Of course, he is not the one who does the turning—a hidden heavy stone collects the force applied when the human demonstrator or the cook turn a heavy handle that winds up a cord around a drum equipped with a ratchet. Obviously "le Petit Bertrand" believes he is the one doing the job because he not only smiles but also moves his head from side to side with obvious pride while turning his little handle. When we were kids, even though we had seen our father wind up the machine and put away the big handle, we liked to believe that the little guy was moving the spit. The irony of the "Petit Bertrand" is that, although the delegation to mechanisms aims at rendering any human turnspit useless, the mechanism is ornamented with a constantly exploited character "working" all day long.

Although this turnspit story offers the opposite case from that of the door closer in terms of figuration (the groom on the door does not look like a groom but really does the same job, whereas "le Petit Bertrand" does look like a groom but is entirely passive), they are similar in terms of delegation (you no longer need to close the door, and the cook no longer has to turn the skewer). The "enunciator" (a general word for the author of a text or for the mechanics who devised the spit) is free to place or not a representation of him or herself in the script (texts or machines). "Le Petit Bertrand" is a delegated version of whoever is responsible for the mechanism. This is exactly the same operation as the one in which I pretended that

Figure 8.3
Le Petit Bertrand is a mechanical meat roaster from the sixteenth century that
ornaments the kitchen of the Hotel-Dieu de Beaune, the hospital where the author
was born. The big handle (bottom right) is the one that allows the humans to
wind up the mechanism; the small handle (top right) is made to allow a little
nonhuman anthropomorphic character to move the whole spit. Although the
movement is prescribed back by the mechanism, since the Petit Bertrand smiles
and turns his head from left to right, it is believed that it is at the origin of the
force. This secondary mechanism—to whom is ascribed the origin of the force—is
unrelated to the primary mechanism, which gathers a large-scale human, a
handle, a stone, a crank, and a brake to regulate the movement.

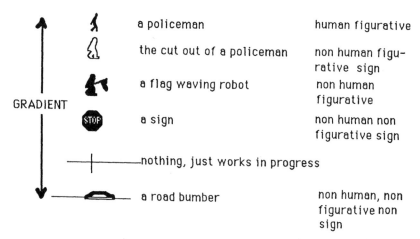

	a policeman	human figurative
	the cut out of a policeman	non human figurative sign
	a flag waving robot	non human figurative
	a sign	non human non figurative sign
	nothing, just works in progress	
	a road bumber	non human, non figurative non sign

Figure 8.4
Students of technology are wary of anthropomorphism that they see as a projection of human characters to mere mechanisms, but mechanisms are another "morphism," a nonfigurative one that can also be applied to humans. The difference between "action" and "behavior" is not a primary, natural one.

the author of this article was a hardcore technologist (when I really am a mere sociologist—which is a second localization of the text, as wrong as the first because really I am a mere philosopher ...). If I say "we the technologists," I propose a picture of the author of the text as surely as if we place "le Petit Bertrand" as the originator of the scene. But it would have been perfectly possible for me and for the mechanics to position *no figurated character* at all as the author *in* the scripts *of* our scripts (in semiotic parlance there would be no *narrator*). I would just have had to say things like "recent developments in sociology of technology have shown that ..." instead of "I," and the mechanics would simply have had to take out "le Petit Bertrand," leaving the beautiful cranks, teeth, ratchets, and wheels to work alone. The point is that removing the "Petit Bertrand" does not turn the mechanism into a "mere mechanism" where no actors are acting. It is just a different choice of style.

The distinctions between humans and nonhumans, embodied or disembodied skills, impersonation or "machination," are less interesting that the complete chain along which competences and actions are distributed. For instance, on the freeway the other day I slowed down because a guy in a yellow suit and red helmet was waving a red flag. Well, the guy's moves were so regular and he was located so dangerously and had such a pale though smiling face that, when I passed by, I recognized it to be a machine (it failed the Turing test, a cognitivist would say). Not only was the red flag delegated; not only was the arm waving the flag also delegated; but the body appearance was also added to the machine. We road engineers (see? I can do it again and carve out another author) could move much further in the direction of figuration, although at a cost: we could have given him electronics eyes to wave only when a car approaches, or have regulated the movement so that it is faster when cars do not obey. We could also have added (why not?) a furious stare or a recognizable face like a mask of Mrs. Thatcher or President Mitterand—which would have certainly slowed drivers very efficiently.[22] But we could also have moved the other way, to a *less* figurative delegation: the flag by itself could have done the job. And why a flag? Why not simply a sign "work in progress?" And why a sign at all? Drivers, if they are circumspect, disciplined, and watchful will see for themselves that there is work in progress and will slow down. But there is another radical, nonfigurative solution: the road bumper, or a speed trap that we call in French "un gendarme couché," a laid policeman. It is impossible for us not to slow down, or else we break our suspension. Depending on where we stand along this chain of delegation, we get classic moral human beings endowed with self-respect and able to speak and obey laws, or we get stubborn and efficient machines and mechanisms; halfway through we get the usual power of signs and symbols. It is the complete chain that makes up the missing masses, not either of its extremities. The paradox of technology is that it is thought to be at one of the extremes, whereas it is the ability of the engineer to travel easily along the whole gradient and substitute one type of delegation for another that is inherent to the job.[23]

	Figurative	Non-Figurative
Human	"I"	"Science shows that"...
Non-Human	"le Petit Bertrand"	a door-closer

Figure 8.5
The distinction between words and things is impossible to make for technology because it is the gradient allowing engineers to shift down—from words to things—or to shift up—from things to signs—that enables them to enforce their programs of actions.

From Nonhumans to Superhumans

The most interesting (and saddest) lesson of the note posted on the door at La Villette is that people are not circumspect, disciplined, and watchful, especially not French drivers doing 180 kilometers an hour on a freeway a rainy Sunday morning when the speed limit is 130 (I inscribe the legal limit in this article because this is about the only place where you could see it printed in black and white; no one else seems to bother, except the mourning families). Well, that is exactly the point of the note: "The groom is on strike, *for God's sake*, keep the door closed." In our societies there are two systems of appeal: nonhuman and superhuman—that is, machines and gods. This note indicates how desperate its anonymous frozen authors were (I have never been able to trace and honor them as they deserved). They first relied on the inner morality and common sense of humans; this failed, the door was always left open. Then they appealed to what we technologists consider the supreme court of appeal, that is, to a nonhuman who regularly and conveniently does the job in place of unfaithful humans; to our shame, we must confess that it also failed after a while, the door was again left open. How poignant their line of thought! They moved up and backward to the oldest and firmest court of appeal there is, there was, and ever will be. If humans and nonhuman have failed, certainly God will not deceive them. I am ashamed to say that when I crossed the hallway this February day, the door *was* open. Do not accuse God, though, because the note did not make a direct appeal; God is not accessible without mediators—the anonymous authors knew their catechisms well—so instead of asking for a direct miracle (God holding the door firmly closed or doing so through the mediation of an angel, as has happened on several occasions, for instance when Saint Peter was delivered from his prison) they appealed to the respect for God in human hearts. This was their mistake. In our secular times, this is no longer enough.

Nothing seems to do the job nowadays of disciplining men and women to close doors in cold weather. It is a similar despair that pushed the road engineer to add a golem to the red flag to force drivers to beware—although the only way to slow French drivers is still a good traffic jam. You seem to need more and more of these figurated delegates, aligned in rows. It is the same with delegates as with drugs; you start with soft ones and end up shooting up. There is an inflation for delegated characters, too. After a while they weaken. In the old days it might have been enough just to have a door for people to know how to close it. But then, the embodied skills somehow disappeared; people had to be reminded of their training. Still, the simple inscription "keep the door closed" might have been sufficient in the good old days. But you know people, they no longer pay attention to the notice and need to be reminded by stronger devices. It is then that you install automatic grooms, since electric shocks are not as acceptable for people as for cows. In the old times, when quality was still good, it might have been enough just to oil it from time to time, but nowadays even automatisms go on strike.

It is not, however, that the movement is always from softer to harder devices, that is, from an autonomous body of knowledge to force through the intermediary situation of worded injunctions, as the La Villette door would suggest. It goes also the other way. It is true that in Paris no driver will respect a sign (for instance, a white or yellow line forbidding parking), nor even a sidewalk (that is a yellow line plus a fifteen centimeter curb); so instead of embodying in the Parisian consciouness an *intrasomatic* skill, authorities prefer to align yet a third delegate (heavy blocks shaped like truncated pyramids and spaced in such a way that cars cannot sneak through); given the results, only a complete two-meter high continuous Great Wall could do the job, and even this might not make the sidewalk safe, given the very poor sealing efficiency of China's Great Wall. So the deskilling thesis appears to be the general case: always go from intrasomatic to *extrasomatic* skills; never rely on undisciplined people, but always on safe, delegated nonhumans. This is far from being the case, even for Parisian drivers. For instance, red lights are usually respected, at least when they are sophisticated enough to integrate traffic flows through sensors; the delegated policemen standing there day and night is respected even though it has no whistles, gloved hands, and body to *enforce* this respect. Imagined collisions with other cars or with the absent police are enough to keep them drivers check. The thought experiment "what would happen if the delegated character was not there" is the same as the one I recommended above to size up its function. The same *incorporation* from written injunction to body skills is at work with car manuals. No one, I guess, casts more than a cursory glance at the manual before starting the engine of an unfamiliar car. There is a large *body* of skills that we have so well embodied or incorporated that the mediations of the written instructions are useless.[24] From extrasomatic, they have become intrasomatic. Incorporation in human or "excorporation" in non-human bodies is also one of the choice left to the designers.

The only way to follow engineers at work is not to look for extra- or intrasomatic delegation, but only at their work of *re-inscription*.[25] The beauty of artifacts is that they take on themselves the contradictory wishes or needs of humans and non-humans. My seat belt is supposed to strap me in firmly in case of accident and thus impose on me the respect of the advice DON'T CRASH THROUGH THE WINDSHIELD, which is itself the translation of the unreachable goal DON'T DRIVE TOO FAST into another less difficult (because it is a more selfish) goal: IF YOU DO DRIVE TOO FAST, AT LEAST DON'T KILL YOURSELF. But accidents are rare, and most of the time the seat belt should not tie me firmly. I need to be able to switch gears or tune my radio. The car seat belt is not like the airplane seat belt buckled only for landing and takeoff and carefully checked by the flight attendants. But if auto engineers invent a seat belt that is completely elastic, it will not be of any use in case of accident. This first contradiction (be firm and be lax) is made more difficult by a second contradiction (you should be able to buckle the belt very fast—if not, no one will wear it—but also unbuckle it very fast, to get out of your crashed car). Who is going to take on all of these contradictory specifications? The seat belt mechanism—if there is no other way to go, for instance, by directly limiting the speed of the engine, or having roads so bad that no one can drive fast on them. The safety engineers have to re-inscribe in the seat belt all of these contradictory usages. They pay a price, of course: the mechanism is *folded* again, rendering it more complicated. The airplane seat belt is childish by comparison with an automobile seat belt. If you study a complicated mechanism without seeing that it re-inscribes contradictory specifications, you offer a dull description, but every piece of an artifact becomes fascinating when you see that every wheel and crank is the possible answer to an objection. The program of action is in practice the answer to an *antiprogram* against which the mechanism braces itself. Looking at the mechanism alone is like watching half the court during a tennis game; it appears as so many meaningless moves. What analysts of artifacts have to do is similar to what we all did when studying scientific texts: we added the other half of the court.[26] The scientific literature looked dull, but when the agonistic field to which it reacts was brought back in, it became as interesting as an opera. The same with seat belts, road bumpers, and grooms.

Texts and Machines

Even if it is now obvious that the missing masses of our society are to be found among the nonhuman mechanisms, it is not clear how they get there and why they are missing from most accounts. This is where the comparison between texts and artifacts that I used so far becomes misleading. There is a crucial distinction between stories and machines, between narrative programs and programs of action, a distinction that explains why machines are so hard to retrieve in our common language. In storytelling, one calls *shifting out* any displacement of a character to another space time, or character. If I tell you "Pasteur entered the Sorbonne amphitheater," I translate the present setting—you and me—and shift it to another space (middle of Paris), another time (mid-nineteenth century), and to other characters (Pasteur and his audience). "I" the enunciator may decide to appear, disappear, or be represented by a narrator who tells the story ("that day, I was sitting on the upper row of the room"); "I" may also decide to position you and any reader inside the story ("had you been there, you would have been convinced by Pasteur's experiments"). There is no limit to the number of shiftings out with which a story may be built. For instance, "I" may well stage a dialogue inside the amphitheater between two characters who are telling a story about what happened at the Académie des Sciences between, say, Pouchet and Milnes-Edwards. In that case, the room becomes the place *from which* narrators shift out to tell a story about the Academy, and they may or not shift *back in* the amphitheater to resume the first story about Pasteur. "I" may also *shift in* the entire series of nested stories to close mine and come back to the situation I started from—you and me. All these displacements are well known in literature departments (Latour 1988b) and make up the craft of talented writers.

No matter how clever and crafted are our novelists, they are no match for engineers. Engineers constantly shift out characters in other spaces and other times, devise positions for human and nonhuman users, break down competences that they then redistribute to many different actors, and build complicated narrative programs and subprograms that are evaluated and judged by their ability to stave off antiprograms. Unfortunately, there are many more literary critics than technologists, and the subtle beauties of technosocial imbroglios escape the attention of the literate public. One of the reasons for this lack of concern may be the peculiar nature of the shifting-out that generates machines and devices. Instead of sending the listener of a story into another world, the technical shifting-out inscribes the words into *another matter*. Instead of allowing the reader of the story to be *at the same time* away (in the story's frame of reference) and here (in an armchair), the technical shifting-out forces the reader to chose *between* frames of reference. Instead of allowing enunciators and enunciatees a sort of simultaneous presence and communion to other actors, techniques allow both to *ignore* the delegated actors and walk away without even feeling their presence. This is the profound meaning of Butler's sentence I placed at the beginning of this chapter: machines are not talking actors, not because they are unable to do so, but because they might have chosen to remain silent to become agreeable to their fellow machines and fellow humans.

To understand this difference in the two directions of shifting out, let us venture once more onto a French freeway; for the umpteenth time I have screamed at my son Robinson, "Don't sit in the middle of the rear seat; if I brake too hard, you're dead." In an auto shop further along the freeway I come across a device *made for* tired-and-angry-parents-driving-cars-with-kids-between-two-and-five (too old for a baby seat and not old enough for a seat belt) and-from-small-families (without other persons to hold them safely) with-cars-with-two-separated-front-seats-and-head-rests. It is a small market, but nicely analyzed by the German manufacturers and, given the price, it surely pays off handsomely. This description of myself and the small category into which I am happy to belong is transcribed in the device—a steel bar with strong attachments connecting the head rests—and in the advertisement on the outside of the box; it is also pre-inscribed in about the only place where I could have realized that I needed it, the freeway. (To be honest and give credit where credit is due, I must say that Antoine Hennion has a similar device in his car, which I had seen the day before, so I really looked for it in the store instead of "coming across" it as I wrongly said; which means that a) there is some truth in studies of dissemination by imitation; b) if I describe this episode in as much detail as the door I will never been able to talk about the work done by the historians of technology at La Villette.) Making a short story already too long, I no longer scream at Robinson, and I no longer try to foolishly stop him with my extended right arm: he firmly holds the bar that protects him against my braking. I have delegated the continuous injunction of my voice and extension of my right arm (with diminishing results, as we know from Feschner's law) to a reinforced, padded, steel bar. Of course, I had to make two detours: one to my wallet, the second to my tool box; 200 francs and five minutes later I had fixed the device (after making sense of the instructions encoded with Japanese ideograms).

We may be able to follow these detours that are characteristic of the technical form of delegation by adapting a linguistic tool. Linguists differentiate the *syntagmatic* dimension of a sentence from the *paradigmatic* aspect. The syntagmatic dimension is the possibility of *associating* more and more words in a grammatically correct sentence: for instance, going from "the barber" to "the barber goes fishing" to the "barber goes fishing with his friend the plumber" is what linguists call moving through the syntagmatic dimension. The number of elements tied together increases, and nevertheless the sentence is still meaningful. The paradigmatic dimension is the possibility, in a sentence of a given length, of *substituting* a word for another while still maintaining a grammatically correct sentence. Thus, going from "the barber goes fishing" to the "plumber goes fishing" to "the butcher goes fishing" is a tantamount to moving through the paradigmatic dimension.[27]

Linguists claim that these two dimensions allow them to describe the system of any language. Of course, for the analysis of artifacts we do not have a structure, and the definition of a grammatically correct expression is meaningless. But if, by substitution, we mean the technical shifting to another *matter*, then the two dimensions become a powerful means of describing the dynamic of an artifact. The

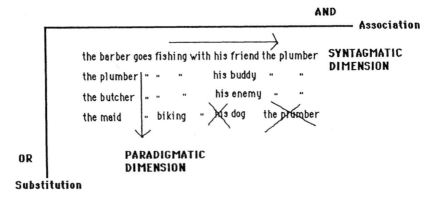

Figure 8.6
Linguists define meaning as the intersection of a horizontal line of association—the syntagm—and a vertical line of substitution—the paradigm. The touchstone in linguistics is the decision made by the competent speaker that a substitution (OR) or an association (AND) is grammatically correct in the language under consideration. For instance, the last sentence is incorrect.

syntagmatic dimension becomes the AND dimension (how many elements are tied together), and the paradigmatic dimension becomes the OR dimension (how many translations are necessary in order to move through the AND dimension). I could not tie Robinson to the order, but through a detour and a translation I now hold together my will and my son.

The detour, plus the translation of words and extended arm into steel, is a shifting out to be sure, but not of the same type as that of a story. The steel bar has now taken over my competence as far as keeping my son at arm's length is concerned. From speech and words and flesh it has become steel and silence and extrasomatic. Whereas a narrative program, no matter how complicated, always remain a text, the program of action substitutes part of its character to other nontextual elements. This divide between text and technology is at the heart of the myth of Frankenstein (Latour 1992). When Victor's monster escape the laboratory in Shelley's novel, is it a metaphor of fictional characters that seem to take up a life of their own? Or is it the metaphor of technical characters that do take up a life of their own because they cease to be texts and become flesh, legs, arms, and movements? The first version is not very interesting because in spite of the novelist's cliché, a semiotic character in a text always needs the reader to offer it an "independant" life. The second version is not very interesting either, because the "autonomous" thrust of a techni-

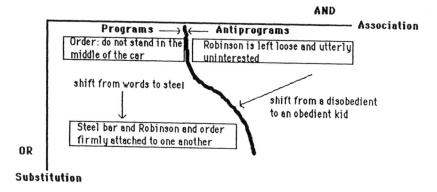

Figure 8.7
The translation diagram allows one to map out the story of a script by following the two dimensions: AND, the association (the latitude so to speak) and OR, the substitution (the longitude). The plot is defined by the line that separates the programs of action chosen for the analysis and the antiprograms. The point of the story is that it is impossible to move in the AND direction without paying the price of the OR dimension, that is, renegotiating the sociotechnical assemblage.

cal artifact is a worn-out commonplace made up by bleeding-heart moralists who have never noticed the throngs of humans necessary to keep a machine alive. No, the beauty of Shelley's myth is that we cannot chose between the two versions: parts of the narrative program are still texts, others are bits of flesh and steel—and this mixture is indeed a rather curious monster.

To bring this chapter to a close and differentiate once again between texts and artifacts, I will take as my final example not a flamboyant Romantic monster but a queer little surrealist one: the Berliner key:[28]

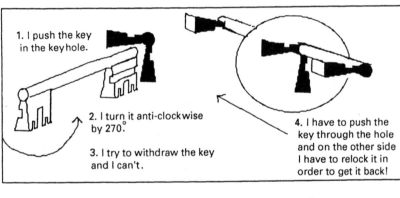

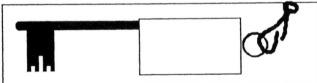

Figure 8.8
The key, its usage, and its holder.

Yes, this is a key and not a surrealist joke (although this is *not* a key, because it is picture and a text about a key). The program of action in Berlin is almost as desperate a plea as in La Villette, but instead of begging CLOSE THE DOOR BEHIND YOU PLEASE it is slightly more ambitious and *orders*: RELOCK THE DOOR BEHIND YOU. Of course the pre-inscription is much narrower: only people endowed with the competence of living in the house can use the door; visitors should ring the doorbell. But even with such a limited group the antiprogram in Berlin is the same as everywhere: undisciplined tenants forget to lock the door behind them. How can you force them to lock it? A normal key[29] endows you with the *competence* of opening the door—it proves you are *persona grata*—but nothing in it entails the *performance* of actually using the key again once you have opened the door and closed it behind you. Should you put up a sign? We know that signs are never forceful enough to catch people's attention for long. Assign a police officer to every doorstep? You could do this in East Berlin, but not in reunited Berlin. Instead, Berliner blacksmiths decided to re-inscribe the program of action in the very shape of the key and its lock—hence this surrealist form. They in effect sunk the contradiction and the lack of discipline of the Berliners in a more "realist" key. The program, once translated, appears innocuous enough: UNLOCK THE DOOR. But here lies the first novelty: it is impossible to remove the key in the normal way; such a move is "proscribed" by the lock. Otherwise you have to break the door, which is hard as well as impolite; the only way to retrieve the key is to push the whole key through the door to the other side—hence its symmetry—but then it is still impossible to retrieve the key. You might give up and leave the key in the lock, but then you lose the competence of the tenant and will never again be able to get in or out. So what do you do? You rotate the key one more turn and, yes, you have in effect relocked the door and then, only then, are you able to retrieve the precious "sesame." This is a clever translation of a possible program relying on morality into a program relying on dire necessity: you might not want to relock the key, but you cannot do otherwise. The distance between morality and force is not as wide as moralists expect; or more exactly, clever engineers have made it smaller. There is a price to pay of course for such a shift away from morality and signs; you have to replace most of the locks in Berlin. The pre-inscription does not stop here however, because you now have the problem of keys that no decent key holder can stack into place because they have no hole. On the contrary, the new sharp key is going to poke holes in your pockets. So the blacksmiths go back to the drawing board and invent specific key holders adapted to the Berliner key!

The key in itself is not enough to fulfill the program of action. Its effects are very severely circumscribed, because it is only when you have a Berliner endowed with the double competence of being a tenant and knowing how to use the surrealist key that the relocking of the door may be enforced. Even such an outcome is not full proof, because a really bad guy may relock the door without closing it! In that case the worst possible antiprogram is in place because the lock stops the door from closing. Every passerby may see the open door and has simply to push it to enter the house. The setup that prescribed a very narrow segment of the human population of Berlin is now so lax that it does not even discriminate against nonhumans. Even a dog knowing nothing about keys, locks, and blacksmiths is now allowed to enter! No artifact is idiot-proof because any artifact is only a portion of a program of action and of the fight necessary to win against many antiprograms.

Students of technology are never faced with people on the one hand and things on the other, they are faced with programs of action, sections of which are endowed to *parts* of humans, while other sections are entrusted to parts of nonhumans. In practice they are faced with the front line of figure 9.2. This is the only thing they can *observe*: how a negotiation to associate dissident elements requires more and more elements to be tied together and more and more shifts to other matters. We are now witnessing in technology studies the same displacement that has happened in science studies during the last ten years. It is not that society and social relations invade the certainty of science or the efficiency of machines. It is that society itself is to be rethought from top to bottom once we add to it the facts and the artifacts that make up large sections of our social ties. What appears in the place of the two ghosts—society and technology—is not simply a hybrid object, a little bit of efficiency and a little bit of sociologizing, but a *sui generis* object: the collective thing, the trajectory of the front line between programs and antiprograms. It is too full of humans to look like the technology of old, but it is too full of nonhumans to look like the social theory of the past. The missing masses are in our traditional social theories, not in the supposedly cold, efficient, and inhuman technologies.

Notes

This paper owes to many discussions held at the Centre de Sociologie de l'Innovation, especially with John Law, the honorary member from Keele, and Madeleine

Akrich. It is particularly indebted to Françoise Bastide, who was still working on these questions of semiotics of technology a few months before her death.

I had no room to incorporate a lengthy dispute with Harry Collins about this article (but see Collins and Yearley 1992, and Callon and Latour, 1992).

Trevor Pinch and John Law kindly corrected the English.

1. The program of action is the set of written instructions that can be substituted by the analyst to any artifact. Now that computers exist, we are able to conceive of a text (a programming language) that is at once words and actions. How to do things with words and then turn words into things is now clear to any programmer. A program of action is thus close to what Pinch et al. (this volume) call "a social technology," except that all techniques may be made to be a program of action. For the technical semiotic vocabulary of this chapter and the next, see the appendix that follows.

2. In spite of the crucial work of Diderot and Marx, careful description of techniques is absent from most classic sociologists—apart from the "impact of technology on society" type of study—and is simply black-boxed in too many economists' accounts. Modern writers like Leroi-Gourhan (1964) are not often used. Contemporary work is only beginning to offer us a more balanced account. For a reader, see MacKenzie and Wacjman 1985; for a good overview of recent developments, see Bijker et al. (1987). A remarkable essay on how to describe artifacts—an iron bridge compared to a Picasso portrait—is offered by Baxandall (1985). For recent essay by a pioneer of the field, see Noble 1984. For a remarkable and hilarious description of a list of artifacts, see Baker 1988.

3. Following Madeleine Akrich's lead (this volume), we will speak only in terms of *scripts* or scenes or scenarios, or setups as John Law says (this volume), played by human or nonhuman actants, which may be either figurative or nonfigurative.

4. After Akrich, I will call the retrieval of the script from the situation *de-scription*. They define actants, endow them with competences, make them do things, and evaluate the sanction of these actions like the *narrative program* of semioticians.

5. Although most of the scripts are in practice silent, either because they are intra- or extrasomatic, the written descriptions are not an artifact of the analyst (technologist, sociologist, or semiotician), because there exist many states of affairs in which they are *explicitly* uttered. The gradient going from intrasomatic to extrasomatic skills through discourse is never fully stabilized and allows many entries revealing the process of translation: user manuals, instruction, demonstration or drilling situations, practical thought experiments ("what would happen if, instead of the red light, a police officer were there"). To this should be added the innovator's workshop, where most of the objects to be devised are still at the stage of *projects* committed to paper ("if we had a device doing this and that, we could then do this and that"); market analysis in which consumers are confronted with the new device; and, naturally, the exotic situation studied by anthropologists in which people faced with a foreign device talk to themselves while trying out various combinations ("what will happen if I attach this lead here to the mains?"). The analyst has to empirically capture these situations to write down the scripts. When none is available, the analyst may still make a thought experiment by comparing presence/absence tables and collating the list of all the actions taken by actors ("if I take this one away, this and that other action will be modified"). There are dangers in such a counterfactual method, as Collins has pointed out (Collins and Yearley 1992), but

it is used here only to outline the semiotics of artifacts. In practice, as Akrich (this volume) shows, the scripts are explicit and accountable.

6. We call the translation of any script from one repertoire to a *more durable* one transcription, inscription, or encoding. This definition does *not* imply that the direction always goes from soft bodies to hard machines, but simply that it goes from a provisional, less reliable one to a longer-lasting, more faithful one. For instance, the embodiment in cultural tradition of the user manual of a car is a transcription, but so is the replacement of a police officer by a traffic light; one goes from machines to bodies, whereas the other goes the opposite way. Specialists of robotics have abandoned the pipe dream of total automation; they learned the hard way that many skills are better delegated to humans than to nonhumans, whereas others may be taken away from incompetent humans.

7. See Authier 1989 on Plutarch's Archimedes.

8. We call prescription whatever a scene presupposes from its *transcribed* actors and authors (this is very much like "role expectation" in sociology, except that it may be inscribed or encoded in the machine). For instance, a Renaissance Italian painting is designed to be viewed from a specific angle of view prescribed by the vanishing lines, exactly like a traffic light expects that its users will watch it from the street and not sideways (French engineers often hide the lights directed toward the side street so as to hide the state of the signals, thus preventing the strong temptation to rush through the crossing at the first hint that the lights are about to be green; this prescription of who is allowed to watch the signal is very frustrating). "User input" in programming language, is another very telling example of this inscription in the automatism of a living character whose behavior is both free and predetermined.

9. In this type of analysis there is no effort to attribute forever certain competences to humans and others to nonhumans. The attention is focused on following how *any* set of competences is *distributed* through various entities.

10. Interestingly enough, the oldest Greek engineering myth, that of Daedalus, is about cleverness, deviousness. "Dedalion" means something that goes away from the main road, like the French word "bricole." In the mythology, science is represented by a straight line and technology by a detour, science by *epistémè* and technology by the *métis*. See the excellent essay of Frontisi-Ducroux (1975) on the semantic field of the name Daedalus.

11. We use *actant* to mean anything that acts and *actor* to mean what is made the source of an action. This is a semiotician's definition that is not limited to humans and has no relation whatsoever to the sociological definition of an actor by opposition to mere behavior. For a semiotician, the act of attributing "inert force" to a hinge or the act of attributing it "personality" are comparable in principle and should be studied symmetrically.

12. I have been able to document a case of a five-day student strike at a French school of management (ESSEC) to urge that a door closer be installed in the student cafeteria to keep the freezing cold outside.

13. It is of course another choice to decide who makes such a choice: a man? a spirit? no one? an automated machine? The *scripter* or designer of all these scripts is itself (himself, herself, themselves) negotiated.

14. This is what Norman (1988) calls the Gulf of Execution. His book is an excellent introduction to the study of the tense relations between inscribed and real users. However, Norman speaks only about dysfunction in the interfaces with the final user and never considers the shaping of the artifact by the engineer themselves.

15. To stay within the same etymological root, we call the way actants (human or nonhuman) tend to extirpate themselves from the prescribed behavior *de-inscription* and the way they accept or happily acquiesce to their lot *subscription*.

16. We call *pre-inscription* all the work that has to be done upstream of the scene and all the things assimilated by an actor (human or nonhuman) before coming to the scene as a user or an author. For instance, how to drive a car is basically pre-inscribed in any (Western) youth years before it comes to passing the driving test; hydraulic pistons were also pre-inscribed for slowly giving back the energy gathered, years before innovators brought them to bear on automated grooms. Engineers can bet on this predetermination when they draw up their prescriptions. This is what is called "articulation work" (Fujimura 1987).

17. We call *circumscription* the organization in the setting of its own limits and of its own demarcation (doors, plugs, hall, introductions).

18. See Suchman for a description of such a setting (1987).

19. We call *ascription* the attribution of an effect to one aspect of the setup. This new decision about attributing efficiency—for instance, to a person's genius, to workers' efforts, to users, to the economy, to technology—is as important as the others, but it is derivative. It is like the opposition between the primary mechanism—who is allied to whom—and the secondary mechanism—whose leadership is recognized—in history of science (Latour 1987).

20. Waddington's term for "necessary paths"—from the Greek *creos* and *odos*.

21. We call *conscription* this mobilization of well-drilled and well-aligned resources to render the behavior of a human or a nonhuman predictable.

22. Trevor Pinch sent me an article from the *Guardian* (2 September 1988) titled "Cardboard coppers cut speeding by third."

A Danish police spokesman said an advantage of the effigies, apart from cutting manpower costs, was that they could stand for long periods undistracted by other calls of duty. Additional assets are understood to be that they cannot claim overtime, be accused of brutality, or get suspended by their chief constable without explanation. "For God's sake, don't tell the Home Office," Mr. Tony Judge, editor of the Police Review Magazine in Britain, said after hearing news of the [Danish] study last night. "We have enough trouble getting sufficient men already." The cut-outs have been placed beside notorious speeding blackspots near the Danish capital. Police said they had yielded "excellent" results. Now they are to be erected at crossings where drivers often jump lights. From time to time, a spokesman added, they would be replaced by real officers.

23. Why did the (automatic) groom go on strike? The answers to this are the same as for the question posed earlier of why no one showed up at La Halle aux Cuirs: it is not because a piece of behavior is prescribed by an inscription that the predetermined characters will show up on time and do the job expected of them. This is true of humans, but it is truer of nonhumans. In this case the hydraulic piston did its job, but not the spring that collaborated with it. Any of the words employed above may be used to describe a setup at any level and not only at the simple one I chose for the sake of clarity. It does not have to be limited to the case where a human deals with a series of nonhuman delegates; it can also be true of relations among non-

humans (yes, you sociologists, there are also relations among things, and *social* relations at that).

24. For the study of user's manual, see Norman 1988 and Boullier, Akrich, and Le Goaziou 1990.

25. Re-inscription is the same thing as inscription or translation or delegation, but seen in its movement. The aim of sociotechnical study is thus to follow the *dynamic* of re-inscription transforming a silent artifact into a *polemical* process. A lovely example of efforts at re-inscription of what was badly pre-inscribed outside of the setting is provided by Orson Welles in *Citizen Kane*, where the hero not only bought a theater for his singing wife to be applauded in, but also bought the journals that were to do the reviews, bought off the art critics themselves, and paid the audience to show up—all to no avail, because the wife eventually quit. Humans and nonhumans are very undisciplined no matter what you do and how many predeterminations you are able to control inside the setting.

For a complete study of this dynamic on a large technical system, see Law (this volume and in preparation) and Latour (forthcoming).

26. The study of scientific text is now a whole industry: see Callon, Law, and Rip 1986 for a technical presentation and Latour 1987 for an introduction.

27. The linguistic meaning of a paradigm is unrelated to the Kuhnian usage of the word. For a complete description of these diagrams, see Latour, Mauguin, and Teil (1992).

28. I am grateful to Berward Joerges for letting me interview his key and his key holder. It alone was worth the trip to Berlin.

29. Keys, locks, and codes are of course a source of marvelous fieldwork for analysts. You may for instance replace the key (excorporation) by a memorized code (incorporation). You may lose both, however, since memory is not necessarily more durable than steel.

9

A Summary of a Convenient Vocabulary for the Semiotics of Human and Nonhuman Assemblies

Madeleine Akrich and Bruno Latour

Semiotics: The study of how meaning is built, but the word "meaning" is taken in its original nontextual and nonlinguistic interpretation; how one privileged trajectory is built, out of an indefinite number of possibilities; in that sense, semiotics is the study of order building or path building and may be applied to settings, machines, bodies, and programming languages as well as texts; the word socio-semiotics is a pleonasm once it is clear that semiotics is not limited to signs; the key aspect of the semiotics of machines is its ability to move from signs to things and back.

Setting: A machine can no more be studied than a human, because what the analyst is faced with are assemblies of humans and nonhuman actants where the competences and performances are distributed; the object of analysis is called a setting or a setup (in French a "dispositif").

Actant: Whatever acts or shifts actions, action itself being defined by a list of performances through trials; from these performances are deduced a set of competences with which the actant is endowed; the fusion point of a metal is a trial through which the strength of an alloy is defined; the bankruptcy of a company is a trial through which the faithfulness of an ally may be defined; an actor is an actant endowed with a character (usually anthropomorphic).

Script, description, inscription, or transcription: The aim of the academic written analysis of a setting is to put on paper the text of what the various actors in the settings are doing to one another; the de-scription, usually by the analyst, is the opposite movement of the in-scription by the engineer, inventor, manufacturer, or designer (or scribe, or scripter to use Barthes's neologism); for instance, the heavy keys of hotels are de-scribed by the following text DO NOT FORGET TO BRING THE KEYS BACK TO THE FRONT DESK, the in-scription being: TRANSLATE the message above by HEAVY WEIGHTS ATTACHED TO KEYS TO FORCE

CLIENTS TO BE REMINDED TO BRING BACK THE KEYS TO THE FRONT DESK. The de-scription is possible only if some extraordinary event—a crisis—modifies the direction of the translation from things back to words and allows the analyst to trace the movement from words to things. These events are usually the following: the exotic or the pedagogic position (we are faced with a new or foreign setup); the breakdown situation (there is a failure that reveals the inner working of the setup); the historical situation (either reconstructed by the analyst through archives, observed in real time by the sociologist, or imagined through a thought experiment by the philosopher); and finally the deliberate experimental breaching (either at the individual or the collective level). No description of a setting is possible or even thinkable without the mediation of a trial; without a trial and a crisis we cannot even decide if there is a setting or not and still less how many parts it contains.

Shifting out, shifting in: Any displacement to another frame of reference that allows an actant to leave the ego. hic. nunc—shifting out—or to come back to the departure point—shifting in. For narratives there are three shiftings: actorial (from "I" to another actor and back), spatial (from here to there and back), temporal (from now to then and back); in the study of settings one has to add a fourth type of shifting, the material shifting through which the matter of the expression is modified (from a sign FASTEN YOUR SEAT BELT, for instance, to an alarm), or from an alarm to an electric link between the buckle and the engine switch, or, conversely, from an electric current to a routinized habit of well-behaved drivers; the first direction is called shifting down (from signs to things) and the other shifting up (from things to signs).

Program of actions: This term is a generalization of the narrative program used to describe texts, but with this crucial difference that any part of the action may be shifted to different matters; if I write in a text that Marguerite tells Faust, "Go to hell," I am shifting to another frame of reference inside the narrative world itself without ever leaving it; if I tell the reader, "go to page 768," I am shifting already away from the narration, laterally so to speak, since I now wait for the reader-in-the-flesh to do the action; if I then write the instruction, "go to line 768," not to a reader but to my computer, I am shifting the matter of the expression still more (machine language, series of 0 and 1, then voltages through chips); I do not count on humans at all to fulfill the action. The aim of the description of a setting is to write down the program of actions and the complete list of substitutions it entails and not only the narrative program that would transform a machine in a text.

Antiprograms: All the programs of actions of actants that are in conflict with the programs chosen as the point of departure of the analysis; what is a program and what is an antiprogram is relative to the chosen observer.

Prescription; proscription; affordances, allowances: What a device allows or forbids from the actors—humans and nonhuman—that it anticipates; it is the morality of a setting both negative (what it prescribes) and positive (what it permits).

Subscription or the opposite, de-inscription: The reaction of the anticipated actants—human and nonhumans—to what is prescribed or proscribed to them; according to their own antiprograms they either underwrite it or try to extract themselves out of it or adjust their behavior or the setting through some negotiations. The gap between the prescriptions and the subscriptions defines the presence or absence of a crisis allowing the setting to be described; if everything runs smoothly, even the very distinction between prescription and what the actor subscribes to is invisible because there is no gap, hence no crisis and no possible description.

Pre-inscription: The competences that can be expected from actors before arriving at the setting that are necessary for the resolution of the crisis between prescription and subscription.

Circumscription: The limits that the setting inscribes in itself between what it can cope with—the arena of the setting—and what it gives up, leaving it to the preinscription. The glass walls of a bar circumscribe the setting; the word "end" at the end of a novel circumscribes the text; the rigid photovoltaic cell kit circumscribes itself and keeps away "idiots" with whom it cannot cope.

Conscription: It is never clear where the "real" limits of a setting are even though it has inscribed precise walls to itself—a book does not end with the word "end" no more than a bar stops at its glass wall; conscription is the series of actors that have to be aligned for a setting to be kept in existence or that have to be aligned to prevent others from invading the setting and interrupting its existence; it is what makes the pre-inscription more favorable for a setting; it is the network effect of any setting, its tendency to proliferate (the book needs librarians, publishers, critics, and paper, and the bar needs whiskey manufacturers, advertising, a heat spell, socializing buddies, etc.)

Interface or plugs: The many gaps between preinscription, circumscription, and conscription are tentatively limited by plugs, sieves, "decompression chambers," or more generally interfaces; when a setting is largely made of materialized interfaces, it looks like a network in the technological meaning of the word: electricity, telephones, water distribution, and sewage systems are peculiar settings that have a network shape.

Re-inscription: The same thing as inscription but seen as a movement, as a feedback mechanism; it is the redistribution of all the other variables in order for a setting to cope with the contradictory demands of many antiprograms; it usually means a complication— a folding—or a sophistication of the setting; or else it means that the complication, the sophistication is shifted away into the pre-inscription; the choices made for the re-inscription defines the drama, the suspense, the emplotment of a setting.

Redistributing competences and performances of actors in a setting: The new point of departure for observation instead of the divide between humans and nonhumans; the directions of this redistribution are many: extrasomatic, intrasomatic; soft-wire, hard-wire; figurative, nonfigurative; linguistic, pragmatic; the designer may shift the competence IS AUTHORIZED TO OPEN THE DOOR either inside a key (excorporation) or inside a memorized code (incorporation); the code itself may be soft-wired or hard-wired (tied to a nursery rhyme, for instance); the task of opening the door may be either shifted to humans or to nonhumans (through the figurative attribution of electronic eyes); the basic competence for opening the door may either be written down through instructions, (linguistic level) as for airplanes, or shifted to the pragmatic level (emergency one-way exit doors that open when pressed upon by a panicked crowd).

A setting is thus a chain of H(umans) and N(onhumans), each endowed with a new competence or delegating its competence to another: in the chain one may recognize aggregates that look like those of traditional social theory: social groups, machines, interface, impact.

Ascription: The attribution process through which the origin of the activity of the setting is finally decided in the setting itself; it is not a primary mechanism like all the others but a secondary one; for instance, the movement of the setting may be ascribed to the autonomous thrust of a machine, to the Stakhanovist courage of workers, to the clever calculations of engineers, to physics, to art, to capitalism, to corporate bodies, to chance, etc.

Scribe, enscripter, scripter, designer, or author: Who or what is the designer of a setting is the result of a process of ascription

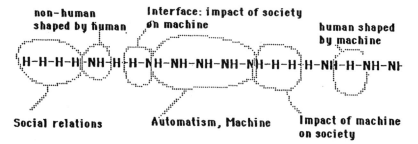

Figure 9.1
The usual categories that sharply divide humans and nonhumans correspond to
an artificial cutting point along association chains. When those are drawn, it is
still possible to recognize the former categories as so many restricted chains. If we
replace H and NH by the name of specific actants, we obtain a syntagm. If we
subsitute a specific name for another, we obtain the shifting paradigms.

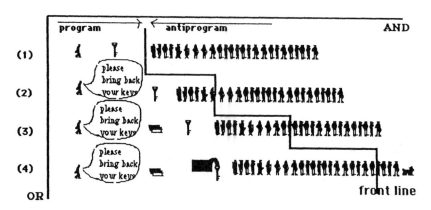

Figure 9.2
The hotel manager successively adds keys, oral notices, written notices, and finally
metal weights; each time he thus modified the attitude of some part of the "hotel
customers" group while he extends the syntagmatic assemblage of elements.

or attribution; but this origin may be inscribed under many guises in the setting itself—trademarks, signatures, legal requirements, proofs that standards are fulfilled, or more generally what the industry calls "traceability"; the blackest of black boxes are illuminated with such inscriptions.

AND (syntagmatic, association, alliances); OR (paradigmatic, substitution, translation): The two fundamental dimensions for following the reinscription of a setting, hence its dynamic or history; the oral or written message BRING YOUR KEY BACK TO THE FRONT DESK is not necessarily obeyed—antiprogram; the shift from keys to weights ties the clients to the front desk because they have a heavy load in their pockets; other antiprograms will appear that will have to be defeated; the front line between programs and antiprograms maps out the plot of a script and keeps track of its history.

10

Technology, Testing, Text: Clinical Budgeting in the U.K. National Health Service

Trevor Pinch, Malcolm Ashmore, and Michael Mulkay

Defining Technology

Technology, unlike science, is everywhere. We use it—to obtain crisp five-pound notes from the automated bank teller; we talk about it—praising the quality of our latest compact disc recording; we write about it—in an attempt to build our careers in the sociology of technology; we construct fantasies around it—such as when one of the editors of this collection drops us at the station in his 1938 Citroen and surprised Dutch people look up to see which movie stars have arrived in town; we may live by it—the dialysis machine; and, we may die by it—the ballistic nuclear missile. As Langdon Winner (1977) remarks, "technology is a word whose time has come."

Providing a definition of something that is so much a part of the fabric of our everyday lives is to offer a hostage to fortune. The editors of *The Social Construction of Technological Systems* (Bijker, Hughes, and Pinch 1987, 3–4) deftly dealt with this problem by refusing to offer an explicit definition. Instead they gave us a series of paradigmatic cases intuitively taken to be technologies. Certainly the artifacts described in that volume—such as bicycles, nuclear missiles, and cooking stoves—would figure on most people's lists as examples of technologies. But we should be careful. Technology like all other terms is indexical—it takes its meaning from its use. Items are classed as technologies for particular purposes. A pertinent example comes from work on gender and technology. Ethnographic studies of technology in the home show that if women are asked to classify which items they consider to be technologies, the home computer will almost certainly be included whereas the cooking stove almost certainly will not.[1] What counts as a technology can itself be contested.

The appeal to intuition works even less well for the object of analysis in this chapter: clinical budgeting systems. These systems provide a new way of distributing financial resources in hospitals so that clinicians can take more responsibility for planning and spending their own budgets. Clinical budgeting has been in operation for over a decade in the Johns Hopkins University Hospital in the United States, and it is currently being introduced into the British National Health Service (NHS). One would not normally think of a budgeting system as a technology. However, clinical budgeting, like many financial systems, is available as software for computers, and a number of software houses are marketing such systems. We will refer to this particular technology as a "social technology." By this term we seek to denote that such a technology has its origins in the social sciences, and that although it may incorporate some material artifacts such as computers, ultimately its purpose is to produce changes in human behavior; in the case of clinical budgeting, the behavior of clinicians.[2]

The rapidly developing field of sociology of technology has until now predominantly focused on material- and machine-based technologies.[3] By describing artifacts and processes whose purpose is to produce changes in human behavior we hope to extend the scope of current work. It is certainly not our intention, however, to once more resurrect old-fashioned distinctions between "hard" material and machine technologies and "soft" social technologies. Indeed, we are confident that the considerations we raise are sufficiently general to apply equally well to all sorts of technology.

We want to show how technologies get described and represented in texts for particular purposes, including that of sociological analysis. Our interest is in the rhetoric[4] of technology. Our specific concern is with the rhetoric through which one technology—clinical budgeting—gets defined, tested, and evaluated within the British hospital system.

Defining Clinical Budgeting
Part of the rhetoric of technology can be exemplified by what seems to be the most trivial of issues—giving a definition of how a clinical budgeting system works. If definitions of technology are contested, then how can we ourselves provide a definitive definition?[5] In an earlier version of this essay we did provide a rather straightforward definition. We wrote,

"Clinical Budgeting" and its close relation, "Management Budgeting" are financial decision-making systems which are intended to give users of health care resources, and in particular clinicians, a greater degree of choice over how resources are allocated such that overall, resources may be used in a more efficient way. (Pinch, Ashmore, and Mulkay 1987, 15)

One year later we find ourselves unhappy with this definition. One particular feature concerns us: the underlying rationale of clinical budgeting, which we described in terms of principles drawn from economics. Specifically, we referred to clinical budgeting as a method for making choices over resource allocation to produce greater "efficiency." This way of describing clinical budgeting is, of course, part of the familiar language with which economists address such issues. Later on in the same paper, we left little doubt that we saw economic principle as being at the core of clinical budgeting:

Clinical Budgeting thus embodies general assumptions concerning economic behaviour which if implemented would change clinicians' behaviour such that they would become economic actors concerned with weighing up costs and benefits rather than merely pursuing treatments regardless of economic consequences. (Pinch, Ashmore, and Mulkay 1987, 19)

Of course, we were not just putting forward an arbitrary definition of clinical budgeting; we had evidence to support our particular definition. For instance, the reason we chose to study clinical budgeting at all stemmed from our ongoing research into health economics. Health economists, when interviewed, had often drawn our attention to the importance of clinical budgeting as a research development in health economics (Ashmore, Mulkay, and Pinch 1989). At a meeting of their leading forum for discussion, the Health Economists' Study Group (HESG), they devoted a special session to a discussion of clinical budgeting.[6] What is not in dispute is that health economists *have* been at the forefront of advocating and implementing clinical budgeting systems and that some of them see clinical budgeting as embodying fundamental principles of economics. However, we would now question whether this is the sole rationale. Perhaps the point can best be made by noting that the persuasive power of health economics itself need not necessarily rest upon an appeal to fundamental economic principle. Let us examine why this might not be so.

In an earlier paper (Mulkay, Pinch, and Ashmore 1987) we examined a series of articles published by two well-known health economists in a leading medical journal widely read by doctors. We found that in these articles two contrasting versions of health economics were presented. In one version, which we called the "strong program," health economics was described as being all about fundamental economic principle. Individuals were treated as rational economic calculators weighing the costs and benefits of their actions in order to maximize their benefits, thereby producing the most efficient overall use of scarce resources. This version of health economics entailed the need for a radical change in what was portrayed as the inefficient and irrational behavior that currently bedeviled the NHS. However, accompanying this "hard sell" of health economics was the "soft sell" or "weak program." In this weak-program version, health economics was presented as "user friendly," as something that could be helpful to health-service practitioners and that involved no radical change in current practices. In the series of articles we examined, the health economists seemed to oscillate between the two versions; sometimes, indeed, using both versions in a single article. As we shall see, this dual rhetoric of health economics is crucial to our analysis.

If health economics itself can be presented in two ways, it is perhaps not surprising that one of the developments intimately associated with health economics in Britain—clinical budgeting—can also be presented in more than one way. The version we gave in our earlier draft, which presents the core of clinical budgeting as being about economic principle, is consonant with the strong program of health economics. However, it is also possible to find a weak-program version.

The following quotations come from a health economist teaching a course for clinicians, one session of which was devoted to management budgeting.[7] The lecturer—a health economist who works for a regional health authority and who is a specialist in the types of budgeting systems under discussion—is describing them to the doctors. To give the flavor of her rhetoric, we have run together a number of separate quotes:[8]

... it's quite specifically patient related, in other words you're looking at the cost of patients, you're looking at the sort of things you can deliver to patients, it makes sense to the consultants, to the doctors, it also makes sense to the nurses, because under management budgeting we actually have a system of ward budgets and consultant budgets. ... If you are in fact quite happy with the way that your service is going ... in my experience—a lot of consultants actually, are happy with what they're doing now and they just want, you know, want to make sure they're not going to get squeezed. But you know they just chug along and maybe in a few years time they'll make some more changes ... most of the changes are at the margins. A great body of your costs are fixed, it is quite difficult to change ... Management budgeting isn't a panacea. Management budgeting is not going to solve your problems. What it is, it's a searchlight on the management problems ...

Throughout her talk there was little reference to management budgeting being about basic economic principle or the need for a radical change in clinicians' behavior on economic grounds. Instead, the budgeting system is presented as something that will probably only have an effect at "the margins"; it is a "searchlight" on management problems.

In contrast, the passage below (which we also quoted in our earlier draft) offers a version of clinical budgeting based on economic principle:

The central plank of clinical budgeting is that if the use of services was charged to a clinician's budget, higher cost services would be reflected in a faster depletion of the budget, forcing consultants and other doctors to choose between a reduced level of activity and a reduced use of resources for each case. This is precisely the model that economists use in examining consumer behaviour in the market place and it is plausible that consumers trade-off costs against perceived benefits in making their market selections. (West 1986, 2)

This quote comes from a paper critical of clinical budgeting written by a health economist, Peter West, and presented at the HESG session on clinical budgeting. Having treated clinical budgeting as essentially an economic topic, West went on to raise all sorts of "technical" objections to it—such as the lack of adequate incentives and the difficulties in calculating costs properly.

Most of the research on clinical budgeting in Britain has been carried out by a specialist research unit known as CASPE (Clinical Accountability Service Planning and Evaluation) and headed by an ex-NHS administrator, Iden Wickings. On occasion, Wickings has presented clinical budgeting to be a matter of economic principle:

All parties can benefit because the Health Authority through its general manager can make real choices about the balance to be struck across all the clinical budgets, and the clinicians can be given extra discretion and thus have an incentive to use their allocated resources more efficiently in the interests of their own clinical service and its patients. In this way optimising the *output* of the NHS, in terms of the quality and quantity of the service provided, becomes a matter of concern to all those involved. (Wickings and Coles 1985, 3)

The language of "incentives," "allocated resources more efficiently," and "optimizing output" draws upon the discourse of economics. This is perhaps hardly surprising, as the quote is taken from a journal concerned with health economics topics. However, in a rather different context—an article in the *Health and Social Service Journal*—Wickings is much less hard-hitting about the need for a change in clinicians' economic behavior. Indeed, he warns of the overenthusiastic adoption of clinical budgeting in the face of doctors' resistance. He points out that ultimately doctors' and clinicians' freedom to decide how and whom they should treat is an important principle of civil liberty. "Clinical freedom often leads to inefficiency, but it is worth it" (Wickings 1983, 467). Clinical and management budgeting are described in terms of accountability and cooperation. There is no appeal to the economic rhetoric of incentives, or maximizing the use of scarce resources:

Both budget types [clinical and management budgeting] allocate resources for the use of consultants, or others with a specific responsibility for managing clinical care programmes ... the intention is that consultants should expect to be answerable for the effectiveness with which they use the resources the health authority provides. Thus both systems need to obtain the willing cooperation of clinicians if they are to be successful. (Wickings 1983, 466)

The terms "clinical budgeting" and "management budgeting" have thus far been used interchangeably. But perhaps it is simply the case that people are talking about two different types of system, and this explains the variety of definitions we have encountered. There are undoubtedly historical differences between clinical and management budgeting. Iden Wickings, who has developed such systems in Britain since the early 1970s, refers to them generically as "clinical budgeting." However, in 1983 a government inquiry into health service management headed by Roy Griffiths[9] recommended the adoption of something called "management budgeting." In our original paper we saw the two systems as essentially the same. Our warrant for this came from Iden Wickings:

The differences between clinical and management budgeting are more to do with the wider spread of management budgeting, and its incorporation of overhead costs, than with any differences in substance. (Wickings and Coles 1985, 3)

But, as Wittgenstein (1974) has pointed out and research in the sociology of science has substantiated for the case of science (e.g., Collins 1985), similarities and differences are not essential properties of objects but are construals placed on objects by our interpretative schemes. For most purposes clinical budgeting and management budgeting can be treated as similar. However, it is possible to find occasions on which the differences are seen as sufficiently important to warrant attention. Such an occasion was the HESG session on clinical budgeting. The following quote is taken from Karin Lowson's response to West's critique:[10]

> I mean his first paragraph in which he says, "Griffiths is the watch word for much of what is happening in the NHS at present. Griffiths notices the lack of clear financial information for management, and as a result management budgeting and clinical budgeting are flavours of the month." I mean we're starting to get into problems already because he seems to use clinical budgeting and management budgeting interchangeably ... and I'm not sure whether he's talking about the same things as I am ...

It can be seen that for this particular occasion clinical budgeting was presented as being in some ways significantly different from management budgeting. Lowson went on to diffuse some of West's critique by characterizing it as obsessed with a technical economic version of clinical budgeting rather than with the practicalities of accountability and the problems of implementing such systems in the NHS.

Before we move on to discuss the testing of these budgeting systems, we will look at one more definition of what management budgeting is meant to achieve. In this case our text is a glossy brochure produced by Price Waterhouse, arguably the world's top accounting firm and management consultancy. After the Griffiths report came down in favor of management budgeting, a number of such companies marketed management budgeting systems and the associated computer software. The interesting feature of the discussion of management budgeting in this text is that the company states outright that the purpose of introducing management budgeting can be to cut costs in the NHS:

> Price Waterhouse, in conjunction with a major international software supplier, Comshare Limited, have formulated an approach to management budgeting which takes into account both the spirit and intentions of the Griffiths recommendations on general management.... The aim of introducing management budgeting is to produce information which can be used to influence the behaviour of clinicians and managers, both individually and corporately. The management objective is to reduce the cost of services, or to provide, in accordance with plans agreed by the Authority, more or better services at the same cost, or at a less than proportional increase in cost. (Price Waterhouse 1986, 3)

This definition differs from others we have encountered—including those we have characterized as "strong"—because it goes further than merely posing efficiency as a goal: it actually suggests that one of the aims may be to cut costs.

Summary

Our recourse to a variety of texts, including an earlier version of this paper, to show how clinical budgeting (and/or management budgeting) can be presented and described in a number of different ways draws attention to an important point about technologies.[11] Technologies are often made available through texts, and the meaning given to a technology through such texts can, as we have seen, vary from context to context (and/or audience to audience).[12] Thus a health economist presenting "the technology" to a group of clinicians who are liable to be quite hostile to its introduction stresses that the technology need not change their practices very much. A health economist wishing to make technical economic criticisms of such systems emphasizes that they are ultimately founded upon economic principle. In response, another health economist stresses that such systems were never meant to be described in this "technical" way and that they are more about "accountability" than efficiency. Wickings, writing in the context of a health economics journal, stresses the economic aspects of clinical budgeting; but in the context of a journal with a much wider readership among NHS personnel, he emphasizes the need for cooperation from doctors. The commercial company's brochure, which is most likely to be read by NHS managers, describes the systems in terms of cost cutting. Lastly, in our paper presented to a variety of academic audiences, we chose to emphasize the fundamental economic rationale underlying clinical budgeting. Our rationale for so doing stemmed from our concerns at the time to develop a definition of social technologies that held that the key element of such technologies was the explicit attempt to change human behavior. By stressing the economic model of behavior behind clinical budgeting, we hoped to demonstrate in a compelling way that this was a paradigmatic case of a social technology.[13]

Our conclusion to this section is thus that technologies can be described and presented in texts for many different purposes, including that of sociological analysis. It is only by close attention to the different discursive contexts in which these definitions are offered and an examination of the rhetoric of technology that we can begin to understand the full richness of its multifaceted and interpretative nature.

Testing Technology

One productive research site in the new sociology of technology has been that of technological testing. Historians of technology have drawn attention to the importance of testing in the development of a technology. For instance, Edward Constant (1980, 1983) in his study of the turbine noted how specific traditions of technological testing emerged along with particular technologies. Methods to determine what a technology can do, and to compare and assess different technical designs, may have to be developed de novo with each technology. The Prony brake for measuring the power of a turbine is an example (Constant 1983). Donald MacKenzie (1989), in his study of the development of the ballistic missile, has also focused on testing. He has shown that technological testing, like experimentation in science, is an interpretative process with test results gaining their meaning and ultimately their validity only within a wider context of technical, social, or even political factors. Thus the result of any test need not, on its own, be held to determine the workability of a technology. Different interpretations of test results can be, and are, offered. For example, MacKenzie documents how for a while representatives of the manned-bomber lobby in the United States argued that "successful" tests of missiles on Pacific islands did not demonstrate the adequate working of the missile technology in the conditions of fighting a real nuclear war. As in science (Pinch 1986), the generalization of any result or set of results to a wider context involves a number of assumptions that are always open to challenge.

The strategic importance of testing in terms of the new sociology of technology is rather akin to the work on experimentation in the sociology of scientific knowledge. For the purpose of showing that scientific knowledge was socially constructed, the view that experimentation provides an unmediated handle on the natural world had to be tackled head-on. Similarly, if we are to get away from the view that a technology possesses a set of features or characteristics that are derived directly from the natural (material) world, we have to address the specifics of how such technical properties get established during the course of testing technologies. As Mulkay (1979) and Pinch and Bijker (1984) have argued, it is not enough to show that different meanings can be given to a technological artifact such as a television set. What has to be shown is how the workability of a television set is itself embedded within a wider context of assumptions, beliefs, actions, and texts—all or any of which are open to challenge in principle, but some of which will remain more closed than others in practice.

Testing Clinical Budgeting

It was with this concern in mind that in the earlier draft of this paper we considered in some detail one particular test of a clinical budgeting system. We took as our example a project started by Iden Wickings in 1979, which tested a clinical budgeting system in three districts of the NHS. This particular project, which had substantial amounts of funding from the DHSS, was strategically important for the widespread adoption of management budgeting in Britain from 1983 onward. This was because the Griffiths inquiry, which recommended the wider adoption of such systems, seems largely to have based its conclusions upon Wickings's research.

Our first task was to establish that a test (or something equivalent) had taken place. Since Wickings himself on various occasions referred to his projects as "tests" or "experiments," this was fairly easy to do. For instance, the results of Wickings's 1979 study were written up in a document entitled "Experiments Using PACTs (Planning Agreements with Clinical Teams) in Southend and Oldham HAs (Health Authorities)." At the start of this report is the following statement:

> The CASPE Research Central Team was established in April 1979 by the DHSS. Amongst other things, it was expected to develop and *test* the use of PACTs. PACTs usually incorporate clinical budgeting. (Wickings et al. 1985, 2, our emphasis)

The term "experiment" was repeatedly used throughout the document, and the value of the project as an experiment has often drawn favorable comment. Indeed, one can see here the importance of the rhetoric of the natural sciences in establishing credibility for a social science project. In an editorial introduction to an article by Wickings and Coles, a leading health economist, Tony Culyer, claimed that the research represented one of the few attempts to carry out "real experiments" on management in the NHS:

> Despite the three recent major organisational changes in the National Health Service the most striking features that continue to characterise its management are the absence of *variety of experimentation* in alternative ways of getting things done. ... This folio reports on what is the one outstanding exception to these deficiencies, some real experiments in offering clinicians budgetary *incentives* to be better managers. Their importance is scarcely to

be underestimated, given the uniqueness of such ordinary experiments in Britain. (Culyer 1985, 1)

In their article following Culyer's editorial, Wickings and Coles also emphasized experimentation and testing.

To further substantiate that we were dealing with phenomena equivalent to experimentation in science, we quoted an excerpt of an interview with Wickings:

OK, we've done a series of projects. The Westminster one was the first that I know of in which we did an experiment ... and it certainly seemed to demonstrate some change ... we tried to achieve the same changes just using costing data [rather than giving the clinicians their own budgets] ... we reported the costs to peer groups with various hypotheses about the high cost groups ... and things like that, and saw nothing for three years despite everybody saying how valuable and important the information was. (Pinch, Ashmore, and Mulkay 1987, p. 21)

We commented,

These early projects are couched in the rhetoric of science.[6] Indeed, this quote might well have come from an interview with a natural scientist. (Ibid.)

We concluded,

The resulting report is replete with the jargon of the scientific report: e.g. "The Experiments in Outline," "Results from the First Phase Experiments" and so on. Experimental rhetoric also surrounds the presentations of the findings. (Ibid.)

We finally quoted the Culyer editorial as yet further evidence that we were dealing with something akin to experimentation in science.

However, footnote 6 of our earlier draft indicates that Wickings himself expressed some reservations about having his work characterized as an "experiment":

6. However, interestingly enough later on in the interview Wickings told us that he did not like the word "experiment" and described his research much more in terms of a learning process. (Ibid., 37)

Other than this footnote, alternative versions of Wickings's research were not pursued in our earlier draft. Having established that a test of clinical budgeting had taken place, we reasoned, we could then go ahead and deconstruct the test and show the real social processes whereby this new technology gained acceptance. In particular, we focused on the role of an evaluation group set up to monitor Wickings's research, which seemed to be instrumental in gaining wider adoption for the technology. The paper followed the pattern of what Woolgar (1983) has called "instrumental irony": we set-up the CASPE project as if it was a hard and fast test of clinical budgeting, and then we deconstructed the results by offering a very different reading of the report.

Before moving on to our deconstruction of the report, we would like first to recover the version of Wickings's work that we hinted at in footnote 6 of our original paper. The relevant interview extracts are reproduced below:[14]

Wickings: I get a bit irritated, people say, you know, "You've been doing this for ten years, and what have you shown?" But in fact each time, we've been trying a different approach and we believe that we've been gradually learned the conditions under which it's likely to be successful.

Interviewer: ... did you see it yourself as an experiment?

Wickings: Yes I mean. Yes we tried, in so far as we could, to set it up so that you would get genuine learning, so I suppose, I don't like the phrase "experimenting," but yes, I don't mind, I suppose ... We worked—sorry I'm sort of stammering really—we certainly saw it as being innovative, and therefore worthwhile if you were going to learn from it, of trying to establish some reasonable sorts of controls, you know and such like, it makes it more complicated and such like. And because of the difficulty of learning from these things and forming balanced judgments there were various ways you ought to evaluate the project ...

Interviewer: Returning to the point about the experiment nature of it. I mean I got the impression when I read the beginning of this report [the CASPE report on PACTs] that these were being set up as kind of like *tests* of the idea, and something riding on these particular events ...

Wickings: Yes, I mean, I think that's probably right, I mean there's a limited number of occasions on which you'll get governmental money to try things out ...

Interviewer (laughing): Sure.

Wickings (laughing): Precisely. Particularly if they're expensive as in many senses this was ... that's what the world's like I'm afraid, it makes it very difficult; I often wish I was injecting rats in cages ... I don't like these words "failure" and "success." You know how these things work don't you?

Interviewer: Yes.

Wickings: How it happens, and that's what I meant, that, the things I felt that we could really be encouraged by, were that it was never rejected ...

It can be seen that a rather different version emerges. Wickings is uncomfortable with referring to the work as "experiments," preferring instead to talk about a "learning process." He also rejects talking about his work in terms of success or failure, instead pointing to "things . . . that we could really be encouraged by." Finally, he hints that the reason that the report is replete with scientific jargon is because it is the only way to obtain the large amounts of financial support needed for such work. It is of interest that this weak-program account of his research is not limited to the interview alone. It is possible to go back to the final report of his research and find the occasional remark supporting this version. For instance, the final chapter of the report is entitled "Lessons Learnt," which is rather unusual for a scientific report. Also in this chapter the language of "encouragement" is used when the authors ask, "What was encouraging in the CASPE projects?" (Wickings et al. 1985, 133). However, the report as a whole does seem to be predominantly written in a scientific style.

The significance of this alternative account of Wickings's research, which can be reconstructed largely from the interview, is that if we had emphasized this way of characterizing his enterprise, we would not have been examining a "test" in the conventional sense at all. It would seem that even what counts as a test of a technology can vary from context to context. Also, our neglect of this weak-program portrayal of Wickings's work has some consequences for our own analysis of clinical budgeting.

Deconstructing the CASPE Research on Clinical Budgeting

We return now to the theme of our original paper. Having shown that we were dealing with an "experiment" or "test," we set about deconstructing the results. This turned out to be a fairly simple task. In much work in the sociology of scientific knowledge carried out in what might be called the "deconstructive mode" (Pinch 1986), what look like definitive experimental results are deconstructed by finding an expert critic with a radically different interpretation of the experiments. In this case we found we had to turn no further than to the CASPE report to provide our deconstruction.

A close reading of the report showed that the "experiment" had been rather unsuccessful, which was surprising in view of all the positive claims made for the research and its influence upon the Griffiths Inquiry. The experiment had had to be abandoned altogether in one of the three districts where a "test" had taken place, and there had been much less progress than expected in the other two districts. Quantitative information on changes in resource use brought about by clinical budgeting showed no firm evidence for any changes in clinicians' practices.

The type of clinical budgeting under "test" involved clinicians and managers collaborating to produce a PACT. Expenditures for different aspects of clinical care were set in advance by these agreements. Clearly such PACTs depended crucially on managers and clinicians being able to work together. In East Birmingham, where Wickings abandoned the research early on, such cooperation on the part of managers was not forthcoming. The difficulties in this district and the slow progress made in the two others (Oldham and Southend) were explained away by some unusual organizational changes that the NHS was undergoing at the time. As the authors of the report put it,

During the five year period of the research a large number of fundamental changes occurred in the reorganization of the NHS. In combination with the more usual factors such as staff changes and selected industrial action they provided an environment within which the research took place and against which the results should be measured. (Wickings et al. 1985, 18)

Local management was faced with continual change. In this unsettled environment perhaps it is not surprising that progress was limited. (Ibid., 24)

The citing of the unusually unfavorable environment as a reason for discounting rather poor results was treated by us in our original paper as an example of a general argumentative strategy used in interpreting the results of technological tests. This strategy rests upon a distinction between the *use* of a technology and the *testing* of a technology. We had in mind the type of arguments that MacKenzie (1989) had documented in the ballistic missile case. The argument works by exploiting the potential similarity or difference between the environment of use and that of testing. In MacKenzie's case, those who argued that the missiles worked posited a sufficient degree of similarity between test and use environments, so that a missile that hit the Pacific-island target with the requisite degree of accuracy could be expected to hit a real target in a nuclear war. Others, however, claimed that the conditions of fighting a real nuclear war were sufficiently different to make it problematic to extrapolate the results from the test environment to the use environment. In the CASPE report, a similar move seemed to have been made when it was claimed that there were unusual circumstances (e.g., NHS reorganization) surrounding the test that prevented legitimate extrapolation of the results to normal use conditions. Our clinical budgeting case thus seemed to exhibit features similar to those in the case examined by MacKenzie, but with a rather different outcome.[15]

Whereas he found *successful* test results being challenged by positing a difference between the environment of testing and the environment of use, we found *unsuccessful* results being ameliorated by a similar argumentative move.

Such an argumentative strategy exploits the fact that the environment can in principle always be cited as a reason for the failure of any experiment (Gilbert and Mulkay 1984; Collins 1985). Thus the Evaluation Group set up to monitor the research could conclude that "despite the major difficulties encountered in the research districts, the Evaluation Group is unanimously of the view that *in principle* the PACTs centred budgeting system has all the right ingredients for improved resource management in the NHS, and it should be given the support needed to ensure its wider dissemination within the service" (quoted in Wickings et al. 1985, 7, our emphasis). There is nothing, of course, to stop the PACT clinical budgeting system from always being successful "in principle." What seemed to be needed was a dose of Popperian philosophy to delimit the conditions under which such tests could be falsified.[16]

Even more damaging in our deconstruction was the failure of any of the quantitative measures to show evidence of changing practices brought about by clinical budgeting (such as changes in spending on drugs, X-rays, and consumables; changes in overall resource use when compared with control districts; and changes in patient management and case mix). We felt that this part of our argument was largely uncontentious because Wickings and his colleagues acknowledged that there had been very few "successes." We quoted from a section of the report where they state,

It sounds perverse, and may indeed be so, to regard the experiments reported here as encouraging rather than disappointing ... What was encouraging in the CASPE projects when so much was not very impressive? (Wickings et al. 1985, 133)

The dramatic tension of our paper came when we listed the "points of encouragement." We set up this list as follows (and here we go back to our earlier draft):

This list appears to be so meager when compared with the original aims of the project that we reproduce it here in full. (Pinch, Ashmore, and Mulkay 1987)

After framing the list in this way, it came as no surprise that, when we presented the paper at conferences, our audiences greeted the list with some amusement. The list is as follows:

(i) the management teams in both Oldham and Southend have continued to invest in staff to support the system;

(ii) much technological development occurred which has since been adopted by the Management Budgeting demonstration districts;

(iii) some (although the minority) of consultants liked and used the available systems and a number of beneficial changes were made [these beneficial changes were that clinicians and managers talked together in a new spirit of cooperation];

(iv) the ward sisters in Southend enjoyed being budget holders;

(v) Mr. Jim Blyth, of the Griffith Inquiry Team, was sufficiently impressed to advocate what he called "Management Budgeting" after his visit to Southend and the systems have substantial similarities;

(vi) perhaps of more significance, the National Evaluation Group were supportive in their interim report;

(vii) although there were only limited signs of "success" there have been even fewer suggestions that the overall thrust was wrong. (Wickings et al. 1985, 133–134)

The point about the ward sisters in Southend enjoying being budget holders was found to be particularly amusing.

Thus, what we accomplished in our analysis was as follows. First we established that a test in the natural-science sense of an experiment had occurred. We then deconstructed this supposed test. Finally, we ironicized the positive results that were obtained by showing that they were "soft" in the sense of impressing important groups of people or arising from vaguely defined factors such as people "working better together." As we wrote after presenting the list of "points of encouragement":

Given the effort put into the project and the notable lack of evidence for any change of clinicians' behaviour on economic grounds even by the authors of the reports' own admission, the above would seem to be a meager harvest. What perhaps is most interesting is that five out of the seven successes listed (i, ii, iv, v, vi) seem to be merely points showing the success of the experiment in convincing important groups that it was a success! (Pinch, Ashmore, and Mulkay 1987, 34)

In view of the ease with which we were able to deconstruct the results, our problem became that of explaining why anyone took the CASPE work seriously in the first place. There seemed little doubt that Wickings's work was viewed as successful by the Griffiths inquiry team and the National Evaluation Group. In the Wickings and Coles article, the authors write,

A more basic method of reaching such agreement has recently been tested in some clinical budgeting experiments. ... In early 1985, an independent Evaluation Group chaired by Professor Buller, the previous Chief Scientist at the DHSS, concluded: "The Evaluation Group is not aware of any other system than PACTs that offers similar interaction between managers and clinicians and notes the adoption of a generally similar format by demonstration districts in the management budgeting programme, at a considerably greater introductory cost ... the Evaluation Group is unanimously of the view that in principle this PACTs centred budgeting system has all the right ingredients for improved resource management in the NHS, and it should be given the support needed to ensure its wider dissemination within the service." (Wickings and Coles 1985, 7)

To explain why the "test" had been seen as such a success, we turned our attention to the role played by the Evaluation Group. But before going on to look at this group, we should perhaps first ask: what if we had taken the weak-program version of Wickings's work as definitive? What would our deconstruction have shown then? The answer is: not very much at all. With the weak-program formulation, there would have been no tests or experiments as such to deconstruct, but only a rather loose exercise in which lessons were learned and points of encouragement sought. Indeed, in regard to this weak program it is not hard to see why the CASPE work could be seen as a success—not solving the problems once and for all, of course, but in the real, messy, capricious social world such problems can never be expected to be completely solved, especially at first. As Wickings told us, he had learned a lot about the conditions necessary to introduce such systems and the quality of management needed to make them work. The fact that users such as nurses were happy with the system is also an important achievement—certainly not one to be scorned. After all, if key NHS personnel cannot be convinced to use them, they will fail. And gaining the support of influential actors such as the Evaluation Group, the Griffiths team, and the DHSS is a real success. Such support reinforces the climate of opinion in which such systems are much more likely to be taken seriously. In short, according to this version of events, rather than being a disaster, significant progress had been achieved.

Under the weak-program version, our distinction between the environments of use and testing also becomes much more problematic. According to the weak program, there is no testing in the natural-science experiment sense, but more a process of learning; thus the environment at the time of the research is an important part of learning how that technology will be used rather than an ad hoc reason to explain away a failed test. The "working" of the sorts of argumentative moves to which we drew attention in our deconstruction is predicated upon a strong-program version of testing clinical budgeting.

Our puzzlement over why the CASPE test was taken to be such a success can also be resolved. If one adopts the weak-program rhetoric, the CASPE research was indeed successful. The puzzle only arises when weak- and strong-program rhetorics are juxtaposed, as in our deconstruction. However, it is not only sociologists who have made use of the two rhetorics; the participants themselves have on various occasions relied upon the rhetorics. The strong-program rhetoric stood the CASPE research in good stead on occasion (in terms of getting it funded, for instance), and Wickings has not published a disclaimer saying that he never carried out any experiments or tests. Scientific rhetoric and the appeal to radical change and economic principle were used when the occasion demanded it. Indeed, one wonders how convinced the Griffiths inquiry team would have been had the research been presented as merely a time-consuming learning experience. The difference between our rhetoric and that of the participants is that for our textual purposes we brought the two rhetorics into opposition to achieve "mutually assured deconstruction"; on the other hand, the participants seem to have used the two rhetorics either separately or side by side to reinforce their arguments. They have managed to avoid the potential deconstruction that the use of the dual rhetoric can entail.

In the last part of the paper we would like to show how this dual rhetoric permeates the activities of the Evaluation Group and that the same process of trading off the one rhetoric against the other occurs there.

Deconstructing the Role of the National Evaluation Group

The Evaluation Group was formed by the DHSS when the CASPE project commenced. It consisted of a number of senior managers in the health service, a professor of health economics, a professor of accounting, a senior clinician, a regional nursing officer, and a regional medical officer, all under the chairmanship of the chief scientist at the DHSS.[17] In choosing to look at the Evaluation Group in our earlier paper, we drew a contrast with experimentation in the natural sciences. We remarked that physicists carrying out experiments did not need an Evaluation Group to tell them what they had found. Furthermore, we noted that the mandate of the Group was very wide indeed, and included dissemination:

The Department (DHSS) recommended that if, in the final analysis the Evaluation Group considered the clinical budgeting research to be of value, it would be essential to widely advertise the results, thereby allowing other districts to adopt a similar management style. The Evaluation Group would therefore have an important role to play in the dissemination of the research results. (Wickings et al. 1985, 7)

In view of their wide remit we saw the Evaluation Group as having a rather significant role in the development of the technology:

Thus the Evaluation Group was not only evaluating the test, but, also making future policy, and furthermore disseminating the results of the tests in the light of that policy. In the context of natural science, it is rather as if the scientists, their funders (e.g. the NSF), and the science media, had all become rolled into one organization invested with the power to determine the future development of this whole area of science. As we shall see, this Evaluation Group came to play a key role in the adoption of the new technology. (Pinch, Ashmore, and Mulkay 1987, 24)

Our argument was that a contentious social technology such as this, which inevitably involved various groups with conflicting interests and which was being introduced in the highly politically charged atmosphere of Thatcherite cutbacks in the NHS, would have a problem of legitimation. In effect, the Evaluation Group helped solved this problem by providing a definitive evaluation of the experiment that could then serve to persuade others that the CASPE project was a good thing. A messy and potentially defeasible set of experiments could be turned into a firm policy edict. We argued that this is essentially what happened. An example of this process was that Wickings and Coles, as we saw above, need not now summarize in detail the far from unambiguous results they had obtained. Instead they could quote the impressive positive comment from the Evaluation Group's interim report.[18] Perhaps of more significance overall, the group's report gave the DHSS a firm legitimating warrant to introduce such budgeting systems on a much wider scale.

When we wrote our earlier paper, we had not talked with any members of the Evaluation Group. Subsequently we contacted two members. The politically charged aspect of the group's work can perhaps best be indicated by noting that both group members made significant comments that they insisted be off the record.

One aspect of the rhetoric of evaluation draws on the supposed *independence* of the evaluating group, which is taken to warrant their evaluative judgments. In other words, the evaluating group is regularly portrayed as having no vested interest in the outcome of the CASPE work, which its members are required to assess impartially. This rhetoric of independent evaluation was consistently adopted in the public record of the Evaluation Group's findings. But in our informal conversations with the two group members a rather different account of the group's activities was presented.

One member, for example, told us that during the period of evaluation it was assumed that the Thatcher government would impose some types of management accountability system whatever happened. With this in mind, this member of the Evaluation Group felt that he was engaged in a damage limitation exercise and that he favored the CASPE work when compared with the more crude cost-cutting and "value for money" ideas of the government. The other evaluator we talked with confirmed on record the role played by pressure from the government:[19]

They suddenly decided, in my view too hurriedly, that we were to report, a sudden decision that we were to report. And I know why that was taken because by that stage the Department, the Government, had decided that they really wanted to move ahead with this, and they thought all this pithering around, this slow development, was really a waste of time and energy. We had to make a big push.

Thus it can be seen that there is a version of the process of evaluation in which the wider context of the politics of use and implementation of the system being evaluated are salient features. The predominant rhetoric of evaluation emphasizing the importance of independence was sometimes evident in our conversations. One evaluator stressed this feature at the outset:

The idea was that we would be independent of the groups that were actually involved in the process. So we weren't really part of the promotion activity. We were independent of it. (Ibid.)

But later he emphasized that independence from Wickings was practically impossible to achieve:

I felt that the evaluation group got too close to Iden Wickings. I mean we were too ... we were almost pushed into a role helping him design his system better by feeding back to him the criticisms that were given ... And I can see that in the interests of health service management that that might be a good idea, but from the point of view of doing strict evaluation, I think we should have been more detached than we actually were ... I won't say captured because we are not easy people to capture ... we got pushed into a role of helping the experiments to work. Rather than evaluating them as they stood ... well that's fair enough but I don't think its quite what an evaluation group should be doing. But on the other hand I think it was better having us there. (Ibid.)

Although we have provided little documentation here, it is clear that a rather different rhetoric of evaluation emerges in our talk with these evaluators compared with that found in the published literature. The politics of implementing clinical budgeting and the practicalities of the evaluation exercise itself are represented in terms of a weak program of evaluation, in contrast to the strong-program version of independence. The two rhetorics of evaluation are normally kept very separate.[20] In the public record the Evaluation Group is presented as an independent, impartial group. Off the record and in informal conversations, a very different rhetoric of commitment and partiality is found. It should be noted, however, that both evaluators felt that they had played a useful role and that more evaluation was needed.

Thus, again we can see the importance of the dual rhetoric in the implementation of this technology. And again one rhetoric can be played off against the other to deconstruct the process of evaluation.

Conclusion

We have tried to display and discuss some of the rhetorics of defining, testing, and evaluating a technology. We have examined the rhetorics used by participants and by ourselves as sociologists. We have identified two broad forms of rhetoric: a strong-program rhetoric that draws on economic principle and carries the promise of radical change—change that can be tested and evaluated in an independent and scientific manner; and a weak-program rhetoric that is sensitive to the complex social and political realties of organizational change, presents clinical budgeting in an mild unthreatening way, views research on clinical budgeting as a slow learning process, and recognizes that technologies are evaluated in a practical and political context.

The difference between the current paper and our earlier draft is that we have laid greater emphasis on the availability of the weak-program rhetoric. By drawing attention to another rhetorical version of the technology under scrutiny, we have added to the argument of the earlier paper. The strong-program version—replete with its rhetoric of testing and experimentation, upon which we placed so much stress earlier—is clearly there. It is put to use by participants and is available for sociological deconstruction. In this revised analysis we have shown that there is also a weak-program version of clinical budgeting, with a rhetoric of learning and sensitivity to the political context in which clinical budgeting is being introduced; this rhetoric is used by participants in certain circumstances and is also available for sociological deconstruction.

Having established that there is a weak-program version of clinical budgeting, including its testing and evaluation, we could have chosen to deconstruct this version, too. This could have been achieved by juxtaposing it with the strong program. We could have asked questions such as: If clinical budgeting is not tested in a rigorous manner, how can we be certain any learning has actually occurred? If it doesn't work in the first place, does it matter how many people become convinced? Isn't talking about a learning process just a charitable way of saying that actually the researcher got it totally wrong the first time? How can any meaningful evaluation be carried out without the total independence of the evaluators? The two rhetorics can thus be counterposed, and each can be used to deconstruct the other.

Nevertheless, our recovery of the weak-program rhetoric in the current paper changes the emphasis of our work. The interesting question we are led to ask is: Why on some occasions can the two versions be presented in texts as deconstructions of each other, whereas on other occasions they can be found cohabiting and even reinforcing one another (in what one health economist jokingly called the "Mr. Nice–Mr. Nasty" strategy)? The effect of deconstruction arises only from our bringing them into opposition for our own textual purposes.[21] Just as we have seen that participants can effectively juggle the two rhetorics to their advantage in their texts, we too can engage in similar rhetorical moves in our texts. Clearly a large part of the success of the clinical budgeting rhetoric has been produced by keeping the two versions apart; they are occasioned rhetorics that vary from context to context for different audiences (Gilbert and Mulkay 1984).

We wish to stress that deconstruction is not only an academic exercise—participants engage in it, too. The presence or absence of deconstruction is at the heart of the issue of textual politics and hence the politics of technology. The obduracy or stability of technologies rest in part on the successful management of textual rhetorics.

In this chapter we have tried to show the workings of some of the rhetoric of technology[22] and of the sociology of technology. Given that sociology can itself be seen as a social technology, it is not surprising that rhetorical moves similar to those used by participants can be found in the texts of sociological analysis.[23]

Notes

Earlier versions of this paper have been presented to a variety of seminars and conferences (School, of Independent Studies, Lancaster University; the annual Society for Social Studies of Science meeting, Pittsburgh; the Science Dynamics Group, Amsterdam University; the International Workshop on the Integration of Social and Historical Studies of Technology, University of Twente; Northwestern University; Conference on Rhetoric of Inquiry/New Sociology of Science, University of Iowa; and CRICT, Brunel University). The paper has benefited considerably from the responses of the audiences at all these presentations. Of particular value has been a set of critical remarks produced by our referee Gerard de Vries. It was de Vries who drew our attention to the way in the earlier version we had taken over a "strong program" rhetoric from the natural sciences. The research undertaken for this paper was funded by the UK ESRC Science Studies–Science Policy Initiative, Grant No. A 332550004.

1. Susan Plummer, private communication. See also Andrea Dahlberg, "Notes on the Household and its Use of Information and Communication Technologies: An Ethnographic Approach," discussion paper presented to the Discourse Analysis and Reflexivity Workshop, Brunel University, April 30–May 1, 1988.

2. A similar social technology designed to produce changes in human behavior and also incorporating material artefacts, was the 'separate system' of prison management (Ignatieff 1978). This technology is described in Pinch, Ashmore, and Mulkay 1987.

3. See, for instance, the technologies discussed in MacKenzie and Wacjman 1985 and Bijker, Hughes, and Pinch 1987. See also the other papers in the present volume.

4. By the term rhetoric we mean a set of systematically used recurring textual features whereby texts gain their persuasive power.

5. Some of the problems of doing "definitional work" are discussed in Ashmore, Mulkay, Pinch, and HESG 1989.

6. This meeting took place at the University of Bath, July 7, 1986.

7. This course took place at Bowness-on-Windermere, Cumbria England, March 17–18, 1986.

8. These extracts come from a talk given by Karin Lowson at the Bowness course, entitled "Management Budgeting," March 18, 1986.

9. Griffiths, formerly managing director of a supermarket chain, was brought in specially by Mrs. Thatcher to introduce more commercial management practices into the NHS. See Griffiths et al. 1983.

10. Karin Lowson, "A Response to Peter West," discussant's remarks at the HESG session on clinical budgeting, University of Bath, July 7, 1986.

11. We are using texts in a wide sense to include lectures and comments made in interviews, as well as written texts.

12. This is another way of referring to the interpretative flexibility of technologies (Pinch and Bijker 1984). See also Akrich (this volume) for a study of how technologies gain different meanings in different social contexts.

13. The recovery of the weak-program version of clinical budgeting does not require any change in our definition of a social technology (at least on this occasion for this paper). In the weak-programme version, clinical budgeting is still being introduced to *change* human behavior—albeit only in a mild and comparatively unthreatening way.

14. The interview with Iden Wickings was carried out at the CASPE Research Unit, King's Fund, London, February 2, 1987.

15. It may be that the use of such an argument to challenge a *successful* test is unique to MacKenzie's case study. Ballistic nuclear missiles are one of the few technologies that are exhaustively tested but that have (thankfully) never been put to use.

16. See K. Popper, *The Logic of Scientific Discovery*, London: Clarendon Press, 1959. Our Popperian rhetoric here mirrors that of participants who can employ different versions of Popper's philosophy to deconstruct the work of fellow scientists. See, M. Mulkay and G. N. Gilbert, "Putting Philosophy to Work: Karl Popper's Influence on Scientific Practice," *Philosophy of the Social Sciences* 11: 389–407, 1981.

17. There were also two DHSS civil servants. As one member of the Evaluation Group told us, their role in drafting the evaluation report was not negligible.

18. This same positive comment was quoted at the start of the CASPE report. Indeed the chronology seems somewhat bizarre in that the Evaluation Group's interim report, was written in April 1985 *before* the research being evaluated had been written up (in December 1985). This enabled the positive evaluation to be cited in the latter report. On the other hand, the former report was specifically designed as an *interim* document.

19. Interview with evaluator, March 11, 1988.

20. However, as noted above, one evaluator started our conversation by stressing the importance of being independent—only to say later that it had been practically impossible to stay independent. It is quite typical in interviews for respondents to begin by giving strong formulations that are later modified under more detailed probing.

21. In our earlier paper we also constructed a "super-strong program" rhetoric of experimentation (based on Popperian falsifiability) against which even the strong program as outlined by the participants appeared weak.

22. Steve Woolgar has recently advocated treating technology as text. Although there are similarities between his program and ours, a difference lies in our choice of the term "rhetoric." We are not just examining the systematic properties of accounts in texts, but how texts on particular occasions gain their persuasive power—hence our focus on rhetoric. See S. Woolgar, "The Turn to Technology in the Social Study of Science," paper presented to the Discourse Analysis and Reflexivity Workshop, Brunel University, April 30–May 1, 1988.

23. Thus one of the concerns of this paper can be interpreted as a form of reflexivity, though perhaps, given our lack of specific attention to the rhetorical strategies employed in the *current* paper, of a weak-program kind. For more full-blooded attempts at, reflexive analysis of science, see Mulkay 1985, Woolgar 1988, and Ashmore 1989. For an attempt to write up our research on clinical budgeting in an unconventional manner, see "Clinical Budgeting: Testing in the Social Sciences—a Five Act Play," which appears as a chapter in Ashmore, Mulkay, and Pinch 1989 and in *Accounting, Organizations and Society* 14, 1989: 271–301.

11

Postscript: Technology, Stability, and Social Theory

John Law and Wiebe E. Bijker

On Heterogeneity and Explanation

All relations should be seen as both social and technical—this is one of the basic themes that runs through the studies in this book. Purely social relations are found only in the imaginations of sociologists, among baboons, or possibly, just possibly, on nudist beaches; and purely technical relations are found only in the wilder reaches of science fiction. This, then, is the postulate of heterogeneity—a postulate suggesting that both social determinism and its mirror image, technological determinism, are flawed. This is because neither the (purely) social nor the (exclusively) technical is determinant in the last instance. Indeed, what we call the social is bound together as much by the technical as by the social. Where there was purity, now there is heterogeneity. Social classes, occupational groups, organizations, professions—all are held in place by intimately linked social and technical means.

But what does this suggest about explanation? Can we have *no* recourse to the commonsense categories of society, technology, agency, and the rest? Several reponses to these questions suggest themselves. Thus it is perfectly possible to elevate the issue to a matter of principle. For instance, in the introduction we mentioned Bloor's (1976) principle of symmetry—the demand that true and false beliefs (or, in the case of technology, both devices that work and those that fail)—should be analyzed in the same terms. On the other hand, we also mentioned Callon's radical (1986a) extension of this principle—his controversial[1] view that the social, the technical, and indeed objects in the natural world should be analyzed in the same terms. Many, perhaps most, English-speaking students of sociotechnology reject this view because it is incompatible with the Wittgensteinian and Winchian (1958) tradition of studying cultures as forms of life: machines, it is argued, cannot possibly create their own culture (Collins and Yearley 1991). Callon and Latour (1991) counter by arguing that it is wrong to privilege humans, that a properly symmetrical analysis will consider relations and interaction without assuming that certain entities—people or their beliefs—are the prime movers of those relations.

It is therefore possible to take a principled epistemological stance on these issues—but it is also possible to avoid doing so. The studies gathered here suggest that despite such differences, there are large areas of overlap and commonality among those committed to the idea that sociotechnology may be seen as a heterogeneous and seamless web.[2] If this is so, then the practical problem is how we might discern patterns and regularities in the sociotechnical, without falling back on the old distinctions between the social, the technical, and the cultural.

One way of thinking about this is to note that if groups and organizations are held in place by mixed social and technical means, we cannot assume that they are stable and unitary. Indeed, they may change or dissolve as those means and their effectiveness changes. Their success or otherwise is a contingent matter, not one of necessity, which means (as we suggested in the introduction) that neither technologies nor social institutions move along inexorable trajectories. Indeed, we have seen Law and Callon make ironic use of the notion of trajectory and stress the uncertain and contingent progress of projects on just these grounds. In a similar mode, Bijker's chapter suggests that innovation does not necessarily precede diffusion: the two may take place simultaneously. The basic point, of course, is that sociotechnical ensembles—facts, artifacts, societies—are interpretively flexible (Pinch and Bijker 1987). Only when the self-evident and unambiguous character of such ensembles has been deconstructed does the quest for the origins of their obduracy become relevant.

But what should be made of this contingency? Does it mean that all is so complicated that description displaces explanation? Is the analysis of sociotechnology restricted to "how" questions? Are questions about why some sociotechnical combinations become obdurate and are institutionalized while others do not simply impossible to tackle because of their complexity? Again, the contributors to this volume offer a variety of views. For instance, Latour very deliberately seeks to elide "how" and "why" questions. Elsewhere (Latour 1988c) he has argued that such constellations as classes, countries, kings, or laboratories should not be treated as the *cause* of subsequent events, but rather as a set of *effects*. In other word, they should be seen as the consequence of a set of heterogeneous operations, strategies, and concatenations. In this view, the job of the investigator is not to discover final causes, for there are no final causes. Rather, it is to unearth these schemes and expose their contingency. There is also a moral point here. Latour assumes that those who are powerful achieve that power by boxing others in, borrowing from them, and misrepresenting them. The object is to uncover these strategies of misrepresentation. In his approach, "why" questions are thus converted into "how" questions.[3]

Another possibility is to press deconstruction still further. Here the investigator takes apart not only the strategies, operations, and concatenations of those under study, but also deconstructs the analogous strategies, operations, and concatenations that generate his or her own account. The point, of course, is that if the coherence and consistency of those under study is the product of discursive or non-discursive methods, then the same is equally true for the univocality of the analyst. There are several possible reasons for attempting this reflexive deconstruction. On the one hand, it is a way of undermining the privilege that attaches, explicitly or otherwise, to the analyst's description. The latter becomes just another account. This can be a particularly effective method for emphasizing the way in which what appears to be a simple phenomenon or object—for instance, a test or an artifact—may be quite differently interpreted by different observers.

On the other hand, it can also be used as a heuristic device. Thus if, as seems possible, both we and those we study use broadly the same methods to achieve a degree of solidity, we may learn about our methods when we study others, and learn about their methods while studying ourselves. These, at any rate, are two of the conclusions that can be drawn from the paper by Pinch, Ashmore, and Mulkay. Thus the authors skillfully show that the "social technology" of clinical budgeting means substantially different things to different people—or indeed to the same people under different circumstances. They also show how two discursive styles, which may be displayed as inconsistent in one account, can be treated as complementary within a strategy that tends to reinforce a technology. A similar conclusion can also be drawn from Bowker's piece on patents, whose utility depended on the use of two registers—one internalist and Whiggish, for deployment in a legal context, and the other externalist and deployed, albeit secretly, in an organizational context.

If the desire to avoid reduction leads authors such as Latour and Pinch et al. to dissolve the distinction between description and explanation, this is not the only possibility. Thus some of the contributors to this volume assume that certain social groups are stable enough to be used as a kind of explanatory scenery, under certain circumstances. The clearest example of this approach is to be found in de la Bruhèze's study of the development of AEC policy toward nuclear waste. In this, the various branches of the AEC are assumed to have relatively stable sets of interests, which in part reflect their existing working practices.[4] Perhaps all empirical studies depend upon some such backdrop—its strategic use might be an example of the way in which new, asymmetrical (if somewhat more local and variable) distinctions may be used to explain the seamless web. After all, even in the reflexive studies, everything cannot be deconstructed simultaneously. However, the contributors to this volume are cautious about the status of the social groups that make up their explanatory scenery. They all, for instance, assume that the actors or groups in question are affected by the unfolding dramas in which they are involved: that at the end they may not be the same as when the story started. In short, such authors assume that the backdrop is a partial function of the events that take place in front of it. The aim, then, is to follow Marx's much-quoted if sexist adage that men make history, but not in circumstances of their own choosing.

As a century of Marxist debate has shown, it is difficult to avoid toppling off the fence in one direction or the other. Typically, either the "circumstances" or the "people" come to dominate explanation. It is difficult to achieve a dialectic in which they are balanced and the way they interact is defined. Indeed, recent discussion in the social analysis of technology addresses precisely this issue.[5] More structurally inclined analysts have argued that the kind of approach exemplified in this volume is excessively actor-oriented and pays insufficient attention to the constraints imposed by structure. This kind of debate is common in sociology (Elias 1978) where, however, its terms reflect an increasing tendency to refuse what Giddens (1984) calls the dualism of structure and agency and instead treat agency, like social relations, as a set of strategically and recursively generated transformational propensities.[6]

The specifics of the sociological jargon are not important here. Actors and structures are *both* products, and they are created and sustained together: to create an actor is also to create a structure, and vice versa. We cannot review the sociological debates here. Nevertheless, the concern with sociotechnical stabilization that runs through this volume is close to—we suggested in the introduction a version of—the problem of securing the social order. Accordingly, we wish to consider the relevance of the work described in the preceding chapters for questions of order, control, and structure. We consider agents and their strategies before turning to the more structural dimensions of obduracy.

On Strategies of Obduracy

Let us start with the observation that much of the time people try to devise arrangements that will outlast their immediate attention. That is, they try to find ways of ensuring that things will stay in one place once those who initiated them have gone away and started to do something else. They also—and this amounts to the same thing— try to find ways of doing things simply (Callon and Latour 1981). The deceptively naive fable offered by Latour illustrates both points. It is simpler to pass through a door than a wall. It is simpler, that is, to delegate the process of creating and closing an opening to an artifact than it is to knock down and rebuild the wall each time— simpler, but not so very simple. The problem is that the delegates have to be kept in place. It is no good delegating tasks to artifacts or people if the effort of making sure that they perform as they should is greater than the original effort. The problem, then, is dual. First, it is necessary to delegate. And second, it is important to find ways of efficiently policing the delegates.

We want to suggest that many, perhaps most, strategies for delegating and policing involve two fundamental moves. First, *a distinction is made between inside and outside* and a set of exchanges between the two is defined and regulated (which amounts to the same thing). And second, those who are outside find themselves compelled to participate in those exchanges: what is produced by the inside, and so the inside itself, becomes what Callon calls an *obligatory point of passage.*

To put it this way is to put it very abstractly. How does this work in practice? Consider a simple example—Akrich's description of the use of photoelectric lighting kits in less-developed countries. These kits, which were sent to Polynesia, were designed—as those who conceived them saw it—to be idiot-proof. The inside of the kits was hermetically sealed from the outside. Possible points of entry were minimized. The designers did not want unauthorized people fiddling around: their plugs were nonstandard; batteries were watertight, and exchanges between the kits and their users, were limited and regulated. A docile user was, as it were, designed to be attached to the kit—a user that the designer assumed would be compelled to use the kit in the approved way because of his or her need for electric light.

Here, then, we see a physical attempt to distinguish between inside and outside and regulate the exchanges between them. And we also see a theory about the needs and resources of users—the notion that they would be compelled to use the lighting kit in the approved manner because they needed the light and did not know enough about electricity to subvert the intentions of the designers. In this way, then, a theory about the behavior of actors—Akrich calls this a script—was built into the artifact. As Akrich indicates, the first of these assumptions was correct, but the second was not. People very soon learned how to subvert the cut-out and obtain "unauthorized" electricity. The script was not played out.

The case of the photoelectric lighting kit is an elementary example: with control of the inside, and a theory about how the outside will react to its products, the actor who seeks to build an institution has some hope of attracting and regulating outsiders. The scientific or technical laboratory offers us another, more sophisticated example of the same strategy at work. Here again an inside is distinguished from an outside. The inside achieves a kind of autonomy, at least for a time, because exchanges between inside and outside are regulated: money and resources are, for instance, exchanged for innovations, or the promise of innovations. But here the inside-outside distinction plays another important role, because the autonomy granted to the laboratory is also temporal. Theories about the environment are not, as in the case of a piece of kit, set in concrete. Rather, they are adaptable. Thus it is often possible to run simulations in the laboratory much faster than it is possible to do in real time.[7] Just as we can run a dozen possibilities through our heads in a second before alighting on the best, so dozens or hundreds of trials and errors can be run in a laboratory before a satisfactory option is found.

Unlike the device itself—for instance, the photoelectric lighting kit—the laboratory is thus a kind of time machine. The photoelectric lighting kit cannot jump forward through time to see whether it is attractive to its users and to check that it is not being "misused." The theory of the environment built into it is either right or wrong. There is no possibility of adaptation. By contrast, theories about the future behavior of the environment created in the laboratory may be explored, tested, and altered. The laboratory, and any other analogous space, has many chances to attract and regulate those who use its products. It also makes its mistakes in private, which means that its credibility is less likely to be undermined in the eyes of outsiders.

But how is the inside distinguished from the outside? How are their exchanges controlled? How are outsiders kept in place? The answers to these questions, as the chapters in this volume suggest, are empirically diverse. So far as the barriers are concerned, the case discussed by Akrich reveals the importance of *physical exclusion*. Here, for instance, there was no way of getting into the batteries, which were sealed. Physical exclusion is also important on a larger scale. Thus industrial companies seek to maintain the security of their research efforts in part by means of walls and chain-link fences.

However, the example of the scientific laboratory points to another important possibility: the ability to scale up and down, which in turn relates to *shifts in materials and media*. Thus the kind of modeling work that we mentioned above operates on objects that are more docile and manipulable than the entities they represent. Thoughts are more docile than people. Drawings, algebraic expressions, and a handful of colored pebbles are more malleable than real dikes. Tons of water can be flooded into a model of the Dutch estuary a hundred times more quickly than the North Sea is able to do this in the real world. Such technologies, which generate echelons of depictions and descriptions of ever-increasing simplicity, homogeneity, and docility, are crucial to many strategies for distinguishing between the inside and the outside.[8]

Such distinctions are, however, reinforced and reproduced by a third set of methods for building barriers. These are *organizational arrangements*, which may be of a legal or quasi-legal basis. Chain-link fences tend, after all, to break down and allow unregulated exchanges between inside and outside unless they are policed. In addition, many metaphorical barriers between inside and outside are inscribed in legal, organizational, discursive, or professional arrangements. Consider, for instance, the Bessemer Steel Association described by Misa. This was a patent-pooling agreement that licensed steel producers in the United States. "Inside" the barrier were all the patents needed to make steel. These might be used by steelmakers outside, in exchange for the payment of appropriate royalties. Steelmakers were drawn to the association because they had no alternative: the apparatus of patent law would have extracted punitive damages had any steelmaker chosen to ignore the patents in question. Accordingly, the Bessemer Steel Association attracted clients. It was a successful arrangement that became an obligatory point of passage for steelmakers.

To create this organizational barrier between inside and outside, the Bessemer Steel Association made strategic use of patents. But as

Bowker shows in his study of the geophysical firm Schlumberger, patents do not stand alone. Though they rest on a distinction between inside and outside, they also help to reproduce these inside/outside divisions. Thus, in a legal context they rest on fictions about priority and the immaculate character of the processes and devices that they purport to describe. At the same time, at least in the case described by Bowker, Schlumberger sought to protect them in court not because it thought they could be turned into an obligatory point of passage, but because for various institutional reasons—primarily delay—it believed that such litigation would give Schlumberger the opportunity to work closely with the oil companies and so entrench itself more firmly in the field.

The Bessemer Steel Association and Schlumberger (if not its patents) were obdurate end points—barriers, or a set of arrangements that distinguished between those who were entitled to sell and those who were obliged to buy. Indeed, much of the process of barrier building has to do precisely with distinguishing between who will be inside and who will be outside. It concerns, that is, the allocation of rights and duties. Often these have to do with rights to speak, or the duty to keep silent—a process that involves disenfranchising those who find themselves on the wrong side of the barrier. To the extent that those outside depend on or have an interest in the product, the product and its producers become an obligatory point of passage.

Thus in the course of his discussion of the proper place for fluorescent lighting in the United States before the Second World War, Bijker touches upon a proposal that all fluorescent light fixtures be certified before sale. More successful, and of greater historical significance, is the example described by de la Bruhèze. Here the question concerned the treatment and storage of nuclear waste in the United States. A number of organizations and divisions had putative rights to speak on this topic, and de la Bruhèze describes the way in which they struggled to impose their own views about the substance of the matter and about those who should have rights to participate in the decisionmaking process. This was a messy bureaucratic battle. However, in the end it led to the creation of a barrier between the inside—those who were competent to speak and make decisions—and the outside—those who were not. In part this was organizational. Different committees were, for instance, empowered with different competences. In part, however, it was professional. Certain experts and specific forms of expertise were enfranchised while others, most notably the general public, were disenfranchised.

De la Bruhèze's study illustrates another feature of the way in which barrier building and the regulation of transactions across the boundary can lead to stabilization. Outsiders may find themselves bound not so much by products created within the boundary and exchanged across it, but by the promise of future products. Thus in the case of the treatment and disposal of nuclear waste, the professionals empowered to investigate and recommend a solution to the problem not only differed among themselves. They also took the view that further research was needed if the problem was to be solved. This, however, was all that was needed to keep the public and outside skeptics in their place. The AEC commissioners, who had the power to decide whether or not to allow the development of nuclear power as a source of energy, were satisfied with the promise of a future solution—even though, two decades later, the difficulties are more intractable than ever.

A similar process is described by Law and Callon in the TSR.2 aircraft project. Like a laboratory, this project attracted clients that granted it resources in exchange for the expectation of a future return. And here again that future return was, at least in some views, not forthcoming. The consequence was that the barriers between inside and outside—the carefully regulated crossing points between the project and its environment—ultimately evaporated, along with the project's clients. Institutionalization was followed by dissolution. An example of greater success is provided by Bijker's case of the fluorescent high-intensity lamp. This was designed by a group of managers to allay the fear of the utility companies that the new lamp would threaten their sales of electricity. The lamp was especially effective because of the promise it entailed: it was not yet possible to make such a lamp, but if at some future point this turned out to be possible, then it would certainly consume a large amount of electricity.

The question of who has a right to speak is important in strategies of stabilization and appears in a number of guises, one related to the issue of interpretive flexibility.[9] As indicated, this is the notion that any object, institution, or process may mean different things to different people. As is clear from a number of studies—for instance, those of Bijker, de la Bruhèze, and Pinch et al. in this volume, and Callon (1980) and MacKenzie (1990a)—what appears as a successful innovation from one perspective may be a failing artifact from another. The example of the contraceptive pill given by Bodewitz et al. (1987) is colorful but to the point: in a recent edition of the Spanish *Pharmacopoeia*, estrogen-progesterone combinations were described as a drug for regulating the menstrual cycle, which had the serious side effect of preventing pregnancy.

If those outside, who are skeptical about an innovation, are to be bound either to that innovation or to the organization from which it emerges, then those who are inside have two main options. Either, as we have seen, they have to disenfranchise the skeptics, or they need to transform the outsiders' perceptions of the innovation, enroll them to the inside, and have them subscribe to that "inside reality." There are several examples of the second option among the case studies. Thus Bijker describes the way in which a "science of seeing," which had to do with subjective perception of artificial illumination, was adapted to conform to the interests of the utilities and used as a tool to persuade the public that there was good sense in trying to create higher intensities of artificial lighting. Again, de la Bruhèze talks of the way in which one of the committees on nuclear waste disposal played a role in "educating" the public about the tractability of the problem. And, finally, as a special technique for both transforming perceptions and disenfranchising skeptics, there is the process of authorized technological testing. Thus, as Pinch et al. remind us, just as what counts as a fact of nature is often ambiguous, so too is what should count as a working technology. The success of a device or process is often a matter for dispute. One way of ensuring that the product is successful is to disenfranchise those who might consider it otherwise. This was the strategy pursued in the case of clinical budgeting. Recognizing that this is a highly controversial "social technology," those responsible for its experimental introduction to the British National Health Service arranged to have the results of their experiments judged by the National Evaluation Group—a committee of high-status professionals whose judgment would, or so it was hoped, carry weight. We witness here, then, the social equivalent of the tradition of testing water turbines described by Constant (1983).

On the Frameworks of Obduracy

We have argued that strategies for realizing obduracy comprise efficient combinations of delegating and policing the delegates. The dialectic of action and structure turns on this double requirement. If the strategies for delegating and controlling are successfully deployed, an institution results, an arrangement is stabilized, a structure emerges. Institutionalization cannot, therefore, be detached from the strategies of actors, but neither can it be reduced to these, because the delegates that an actor seeks to array and hold in place are drawn from a structured environment. That structure, like actors or institutions, may be seen as a contingent set of heterogeneous relations. From the standpoint of any particular actor, the structure and the actors defined within it represent a more or less accurately pictured geography of enablement and constraint. Thus, some relations are much easier to create and maintain than others. They are ready to be drawn on and can be utilized simply and economically. Others are expensive, awkward, and time consuming. Structure, then, is something like a system of transport. The network of paths, tracks, roads, railway, and airlines mean that it is easy to get from some places to others. They are close, either figuratively or literally. On the other hand, other locations are far removed from one another. Maintaining links between them is time consuming, tedious, expensive, or downright impossible.

If the relations that make up structure are an emergent consequence of actors' strategies and unmotivated actions and events, then structure is liable to change in ways that are sometimes unpredictable. However, any particular agent can only hope to act in a way that has more than a random chance of success, if the geography of structural relations displays some degree of predictability.[10] We have touched on one of the consequences in an earlier section: for certain purposes, even those who insist on the contingency of structure are able to treat it in practice as a more or less invariant scenery that shapes, but is relatively unshaped by, the action that takes place on the stage. Thus the notion that certain agencies have locally stable interests or practices finds its way into the accounts of a number of our contributors. Accordingly, though most of the authors are at pains to argue that such interests are subject to change,[11] they tend to work on the assumption that actors have a (relatively stable) concern to preserve the structure of their existing practice. This is the backdrop to which we referred earlier.

This is not to say that actors are always, or even typically, aware of the structures within which they operate. Thus, though they have procedures and technologies for ordering and *representing* those structures—for instance, the model of the Dutch estuary system—such procedures are necessarily precarious. This is because they rest on a series of simplificatory assumptions—about the general character of the environment, how it is organized, and how it might be ordered or reordered. As writers from Simon (1969) onward have argued, simplification is a dangerous necessity, for there is no way of representing and handling complexity or nuance in full. Accordingly, such assumptions may or may not turn out to be workable in practice next time around. Thus Carlson's chapter argues that Edison was acting within a specific simplificatory frame—that of producer culture—and that the character of this frame explains why Edison and his associates did not successfully participate in the growth of the mass movie industry. History had, as it were, moved on, and unlike the case of Schlumberger described by Bowker, or the Bessemer Association described by Misa, Edison's strategies did not directly shape the course of that history. Rather, it was the "bottom-up" entrepreneurs opening the nickelodeons in working-class towns who succeeded in operating in that part of the environment. Here, then, Carlson questions the model of calculative rationality that so often, albeit implicitly, underlies the analysis of sociotechnology. According to that model, some calculation of an actor's interests may explain subsequent events. The case of Edison's involvement in the motion picture industry suggests that such a model is at best incomplete and in some cases simply wrong.

If we want to eschew reductionism, what then can be said about the geography of constraint and enablement that makes up the environment? What can be said about the way in which this affects the success of actors' strategies? And what can be said about the circumstances that lead particular concatenations of sociotechnical elements to display particular obduracy in the face of their environments? We want to conclude by pointing to three lines of work that offer possible answers to these questions: first, the notion of *technological frame* as developed by Bijker; second, the notion of *technological momentum* developed by a number of social historians of science; and third (like the first, strongly represented in this volume), the distinction between inside and outside, which leads to the formation of what Law and Callon call *negotiation spaces*.

The notion of technological frame (see Bijker 1987) refers to the concepts, techniques, and resources used in a community—any community, not simply a community of technologists. Technological frame is thus a combination of the explicit theory, tacit knowledge, general engineering practice, cultural values, prescribed testing procedures, devices, material networks, and systems used in a community. It is—and this is what distinguishes it from such possible social analogues as Mary Douglas's (1973) notion of grid and group or Joseph Ben-David's (1960) concept of role hybridization—simultaneously social *and* technical. Actors' meanings, including those parts of their strategies that are explicitly articulated—the ways in which they react to and interpret structure—form a part of technological frame. But so, too, do relations of which the actors are not aware—relations that may be embodied, as in the case of skills, or form part of their environment, as in the case of such resources as the power supply or the details of software that they use to build their spreadsheets. Technological frame is thus concerned with structuring relations, whether social or technical. It is also a bridge between structure and action. And that bridge both points to ways in which structure may be influenced by action and makes it possible to predict that certain kinds of structure will lead to one kind of action, and other structures to alternative actions.

As an example of the way in which action may influence technological frame, consider the case of celluloid. Bijker (1987) describes the way in which the specific attempts of Hyatt, his collaborators, and his competitors to develop nitrocellulose plastics (such as the focus on solvents and in particular camphor as a key element in the invention of celluloid) had a direct impact on the technological frame of the next generation or celluloid engineers. As a result, the courses of action of the chemists subsequently working within that frame were further constrained—but also, of course, enabled. But the theory of technological frame also makes predictions about the style and origin of innovations. Thus under certain circumstances there will be one dominant group that is able to insist upon its definition of both the problems and the appropriate solution. Under such monopolistic circumstances, *conventional* innovations tend to arise. In particular, they do so when there is functional failure (Constant 1980), and they are judged in terms of their perceived adequacy in solving such failures.

Under other circumstances, when there are two or more entrenched groups with competing technological frames, arguments that carry weight in one frame will carry little weight in the other. Under such circumstances criteria external to the frame in question may become important as appeals made to third parties, over the heads of the other social group. In addition, innovations that allow the amalgamation of the vested interests of *both* groups will be sought. Such innovations (the definition of steel and associated testing technologies present a case in point) are, so to speak, doubly conventional because they have to lodge within both technological frames.

The third situation considered by Bijker (and here his case is the early history of the bicycle) occurs when there is no single dominant group and, as a result, no effective set of vested interests. Under such circumstances, if the necessary resources are available to a range of actors, there will be many different innovations. Furthermore, these innovations may be quite radical. More than in the other cases, the success of an innovation depends on the formation of a constituency, a group that comes to adopt the proposed technological frame.[12]

Bijker explains action by relating it to the way in which actors are shaped by and implicated in a network of relations. There are commitments, explicit or otherwise, to economic investments, normal practice, and skills. There is dependence—which is not remarked upon until things start to go wrong—on networks of resources that enable certain courses of action while more or less frustrating others. And there is the question of the differential availability of those networks of resources. Thus it was far more expensive to enter the electricity supply business when it started than to initiate the manufacture of bicycles. The result is a model not of the interests or commitments of specific social groups but of the *patterns* that arise when social groups are constituted and interact with one another in a range of different structural circumstances. It is, in other words, a predictive structural theory about the obduracy or certain sociotechnical circumstances and the malleability of others. It is, moreover, a theory that is neither socially nor technologically reductionist: the concept of "technological frame" is intrinsically heterogeneous.

Though we have mentioned the way in which Law and Callon ironicize the notion of technological trajectory, the concepts of *technological momentum* and the closely related notion of *life cycle* have been deployed with considerable success in the history of technology (see Hughes 1983, Staudenmaier 1985). The argument of such historians is that, at least for America between the 1880s and the 1930s, certain technologies—the cited cases are electricity supply and the motor car—and their carriers, which were malleable in their early stages, later developed to a point at which they were relatively insensitive to, but exercised great influence over, their environments.

Though it is possible that such analyses are historically contingent—they apply to the United States at a particular time, but cannot be applied elsewhere—it is nevertheless interesting to note that on the basis of his theoretical generalizations, Hughes (1986b) has made predictions about the development of the modern health care system. Hughes argues, for example, that at present health care is at a stage of development comparable to that of power systems between the two world wars. As with power systems at that earlier time, the medical systems' components are now heavily capitalized and institutionalized. Hence, the era has passed when independent inventors—for instance, physicians—with limited capital and institutional support could dominate research and development. This suggests that institutions with easy access to capital will take on key roles in the process of building systems. Obvious candidates for such a role are the medical equipment manufacturers, the pharmaceutical companies, and various multipurpose consulting firms and holdings. Drawing on an analogy between pharmaceutical firms in the health care system and petroleum companies in the electric power supply, Hughes predicts that the pharmaceutical companies will only be able to assume a central role in the health care system if they develop a holistic approach to medical problems. Otherwise they may find themselves on the periphery, as were the petroleum companies that were so involved in the automobile industry that integration into the power supply system was difficult.

It is interesting to note that such theories draw on similar intellectual roots as the theory of technological frame. That is, once again they rest on the extent to which actors are shaped by or otherwise implicated in particular networks of relations. Some of these are economic, hence Bijker's use of a vocabulary of investment when he talks of the "amortization" of vested interests.[13] Others take the form of commitments to expertise and embodied skills: the metaphorical investment of time and energy. In addition, however, there are patterned relations—for instance, the highway system in the United States, the character of public transport, the growth of new styles of consumption (such as out-of-town merchandising)—that depend on the maintenance (in this instance) of the automobile. Such patterned relations—what Staudenmaier calls the "maintenance constituency" —add to the obduracy or momentum of the sociotechnical system because they rest on an endless series of "side bets" (Becker 1964). This, however, is a contingency. If the side bets are lost, or reshaped, then the sociotechnology will be accordingly reshaped.[14]

The third approach to middle-range analysis of the way in which structure relates to action—an investigation of the distinction between inside and outside—takes us in rather a different direction. In the previous section we described the way in which strategies of obduracy frequently, if not always, turn first around the creation of a distinction between inside and outside, and second upon ensuring that whatever is inside becomes an obligatory point of passage for those on the outside. We mentioned the simple case of physical exclusion—the paradigm case was the battery intended for a user in the developing world—but also talked about the arrangements that allow those inside the barriers to turn themselves into a kind of time machine and so model the outside. In this context physical, legal, rhetorical, bureaucratic, and technological methods were all mentioned, and doubtless there are many others. The Bessemer patent pool is another example. Creating the patent pool in 1866 was obviously a strategy designed to close the controversy between the Troy and the Ward groups. But it also led to structural constraints to future actions. Thus, as Misa shows, it posed a serious barrier to steelmaking firms, such as Andrew Carnegie's, which were trying to extend their market share in the 1870s.

The inside/outside division is heterogeneous in character. It has to do with the organization of bureaucracies (Chandler 1977), the development of methods of accountancy (ibid.; McGaw 1986; Mac-Kenzie 1990b; Law 1991a), technologies of communication (Eisenstein 1979; Beniger 1986), techniques of representation (Bertin 1983; Latour 1990; Lynch and Woolgar 1990; Tufte 1983, Shapin and Schaffer 1985), methods of modeling (Law 1991b), mathematical tools and statistical representations (MacKenzie 1978, 1990c), developments in cartography (Wilford 1981), legal innovations (Pool 1983), and a host of other sociotechnologies. If the social is too weak to hold us all together (Callon and Latour 1981; Latour 1990), then it is certainly too weak to create obdurate negotiation spaces that are able to model and shape what goes on in the environment.

As is also obvious, such methods do not stand outside history. Rather, like all the other sets of relations that we have touched on, they are historically contingent. We hesitate to make the simple-minded argument that they are in a continuous process of development. Perhaps, like Mann (1986) and Beniger (1986), it would be better to say that they evolve discontinuously. Nevertheless, it is incontrovertible that they are subject to secular change. Such changes are not readily visible in the case studies described in this book. Nevertheless, when the new sociotechnology starts to address these changes, it will begin to obtain purchase on some of the fundamental historical and sociological questions about power, class, inequality, social change, and the formation of the modern world.

Conclusion

To conclude, we return to our point of departure. Technology is never purely technological: it is also social. The social is never purely social: it is also technological. This is something easy to say but difficult to work with. So much of our language and so many of our practices reflect a determined, culturally ingrained propensity to treat the two as if they were quite separate from one another. The authors in this volume all wrestle with this problem. Of course, they do not come to identical conclusions. Their work has a range of contrasting implications for historiography, for social and political theory, and for the organization and management of technical change. What brings them together is an urgent sense of the need to understand the heterogeneous webs in which we are implicated.

We want to conclude with two thoughts. The first is that the academic time is right for work on the sociotechnical. We rest our case on the various approaches exemplified in this book. Of course, they are underdeveloped. Of course, they represent work in progress. Of course, they have limited applicability at present. Nevertheless, we believe that they show how theoretically informed empirical work may start to break down the disciplinary barriers and the habits of common sense and make it possible to understand the sociotechnical world in which we are caught up. This, then, is our first thought: we are witnessing the birth of a new capacity to understand, in a matter-of-fact way, how it is that people and machines work together, how they shape one another, how they hold one another in place.

Our second thought has to do with the urgency of this task. Our technologies surround us, as they have for millennia, but never before have they been so powerful. Never before have they brought so many benefits. Never before have they had such potential for destruction—in many cases a potential that has been realized. And never before has the task of understanding those technologies—how they are shaped, how they shape us—been so urgent. The work described in this book is only a first step. But with its stress on heterogeneity it *is* a first step: it says, in effect, that technical questions are never narrowly technical, just as social problems are not narrowly social. When things go wrong, it may not not make much sense to blame technologies. Neither does it necessarily make sense to blame people, nor even the economic systems in which they are caught up. Who or what should be blamed for the Nimitz Highway collapse? Or the Challenger disaster? Or the deforestation of the Himalayas? Or the greenhouse effect? If we want to make sense of these horrors—and more important, do something about them—it does not really help to look for a scapegoat. Rather, what we urgently need is a tool kit—or rather a series of tool kits—for going beyond the immediate scapegoats and starting to grapple with and understand the characteristics of heterogeneous systems.

Notes

1. See, for instance, Amsterdamska 1990, Collins and Yearley 1991, and Callon and Latour 1991.

2. For a discussion of the metaphor of the seamless web, see Hughes 1986a.

3. Law (1991a), though sympathetic with Latour's moral and methodological position, argues that "how" and "why" questions are not mutually exclusive. Specifically, he suggests that power is indeed the product of a set of (strategy-dependent) relations, but this does not mean that it cannot be stored and used for certain purposes.

4. It could be argued that similar assumptions underlie Misa's chapter on the development of steelmaking in the United States in the nineteenth century, Bijker's study of fluorescent lighting, and Law and Callon's description of an unsuccessful military aircraft project.

5. See, for instance, Russell 1986 and Pinch and Bijker 1986.

6. See Giddens's work—for instance, Giddens 1984; and for an interesting recent commentary, see Clegg 1989. See also Law 1991a.

7. Or if not more quickly, at least more tractably. A concrete example of this is given by Latour (1987, 230–232) when he describes the scale modeling of the Dutch coast undertaken by civil engineers in the Delft Hydraulics Laboratory. Here time in the laboratory is scaled up (tides come in every twelve minutes) and size is scaled down (the whole of the Dutch river estuary is reproduced in one big laboratory hall). Law (1991c) describes a similar strategy for the case of aeroengine design. However, when other sociotechnical ensembles are used to achieve control, the scaling up and down may operate in the other direction. For example, in developing micro-automation technologies, it is often useful to build a laboratory model that is slower and larger than the final product.

8. This strategy also formed the core of Boyle's successful attempt to create a boundary between the inside and outside of science, as analyzed by Shapin and Schaffer (1985). The "material technology" of the air pump delineated the inside of science by defining the terms in which durable knowledge could be stated. Like the Dutch estuary model 300 years later, it defined the distinction between what were to be facts and what not.

9. See Collins 1981b, and Pinch and Bijker 1987.

10. For a careful analysis of structure and power as a distribution of knowledge, see Barnes 1988. Barnes does not, however, consider the role of technology in maintaining relations. For an initial attempt at this, see Law 1991a.

11. This argument may be mounted both for interests imputed by agents to themselves and for those imputed by others, including analysts. Roughly speaking, in both cases interests appear to be predictive attributes that function to link prospective structural features with the set of existing relations that constitute the actor.

12. See Staudenmaier's (1985) discussion of the notion of "constituency."

13. Misa and Bijker talk in this volume of the "amalgamation" of vested interests—a phrase with less restricted economic connotations.

14. For instance, the radical Conservative government of Mrs. Thatcher substantially altered the structure of the side bets of the British electricity supply industry.

This can be seen as a sociotechnical experiment on a huge scale. Its outcome—particularly in terms of patterns of investment in the generation of power—is unclear and is likely to remain so for some time. Some, including its critics, suggest that such investment decisions will henceforth be made on more local, short-term accounting grounds. Whether this actually happens remains to be seen. If, however, it indeed turns out to be the case, then Hughes's predictions about the holistic character of successful participation in sociotechnical systems will be incorrect—at least in this case. This is because the structure of side bets will have been radically altered. Of course, critics of the experiment might argue that Hughes is really right because the lack of holism encouraged by the introduction of local market considerations into electricity supply means that the future security of power supplies, and the overall long-term efficiency of the system, are both put at risk by the entry of a large number of players who calculate in terms of relatively short-term economic considerations. However, the jury is out, and is likely to stay out on this one until well into the next century!

References

Allaud, Louis, and Maurice Martin. 1976. *Schlumberger. Historie d'une Technique.* Paris: Berger-Levrault.

Allen, Robert C. 1982a. "Vitascope/Cinematographe: Initial Patterns of American Film Industrial Practice," in *The American Movie Industry: The Business of Motion Pictures*, G. Kindem, ed. Carbondale, IL: Southern Illinois University Press, 3–11.

Allen, Robert C. 1982b. "Motion Picture Exhibition in Manhattan, 1906–1912: Beyond the Nickelodeon," in *The American Movie Industry: The Business of Motion Pictures*, G. Kindem, ed. Carbondale, IL: Southern Illinois University Press, 12–24.

Amsterdamska, Olga. 1990. "Surely You're Joking Monsieur Latour!" *Science, Technology and Human Values*: 495–504.

Andersen, Håkon With, and John Peter Collett. 1989. *Anchor and Balance: Det Norske Veritas, 1864–1989.* Oslo: J. W. Cappelens Forlag.

Anderson, Robert. 1985. "The Motion Picture Patents Company: A Reevaluation," in *The American Film Industry*, rev. ed., T. Balio, ed. Madison, WI: University of Wisconsin Press, 133–152.

Ashmore, Malcolm. 1989. *The Reflexive Thesis: Wrighting the Sociology of Scientific Knowledge.* Chicago: University of Chicago Press.

Ashmore, Malcolm, Mulkay, Michael and Pinch, Trevor J. 1989. *Health and Efficiency: A Sociology of Health Economics.* Milton Keynes: Open University Press.

Ashmore, Malcolm, Mulkay, Michael, Pinch, Trevor, and the Health Economists' Study Group HESG (1989) "Definitional Work in Applied Social Science: Collaborative Analysis in Health Economics and Sociology of Science," in *Knowledge and Society: Studies in the Sociology of Science Past and Present*, L. Hargens, R. A. Jones, and A. Pickering, eds. vol. 8, Greenwich, Connecticut: JAI Press, 27–55.

Authier, M. 1989. "Archimède, le canon du savant," in *Eléments d'Histoire des Sciences*, Michel Serres, ed. Paris: Bordas, 101–127.

Balio, T. 1985. "Struggles for Control," in *The American Film Industry*, rev. ed., T. Balio, ed. Madison, WI: University of Wisconsin Press, 103–131.

Baker, N. 1988. *The Mezzanine.* New York: Weidenfeld and Nicholson.

Balogh, B. 1987. *Trouble in Paradise. Institutional Expertise in the Development of Nuclear Power, 1945–1975.* Ph.D. diss., Johns Hopkins University.

Barnes, Barry. *T. S. Kuhn and Social Science.* New York: Columbia University Press.

Barnes, Barry. 1988. *The Nature of Power*. Cambridge: Polity Press.

Baxandall, Michael. 1985. *Patterns of Intention. On the Historical Explanation of Pictures*. New Haven, CT: Yale University Press.

Bealer, Alex W. 1969. *The Art of Blacksmithing*. New York: Funk and Wagnalls.

Beamont, Roland. 1968. *Phoenix into Ashes*. London: William Kimber.

Beamont, Roland. 1980. *Testing Years*. London: Ian Allen.

Becker, Howard S. 1964. "Personal Change in Adult Life." *Sociometry* 27: 40–53.

Ben-David, Joseph. 1960. "Roles and Innovations in Medicine." *American Journal of Sociology* 65: 557–568.

Beniger, James R. 1986. *The Control Revolution: Technological and Economic Origins of the Information Society*. Cambridge, MA: Harvard University Press.

Bertin, Jacques. 1983. *Semiology of Graphics: Diagrams, Networks, Maps*. Madison, W: University of Wisconsin Press.

Bessemer, Henry. 1896. "The Bessemer Process." *Engineering* 6 (20 March): 367–370.

Bessemer, Henry. 1905. *Sir Henry Bessemer, F.R.S.: An Autobiography*. London: Offices of Engineering.

Bijker, Wiebe E. 1987. "The Social Construction of Bakelite: Toward a Theory of Invention," in *The Social Construction of Technological Systems*, W. E. Bijker, T. P. Hughes, and T. J. Pinch, eds. Cambridge, MA: MIT Press, 159–187.

Bijker, Wiebe E., Hughes, Thomas P., and Pinch, Trevor J. 1987b. "General Introduction," and "Introductions," in Bijker et al., eds. 1987a, 1–6, 9–15, 107–110, 191–194, 307–309.

Bijker, Wiebe E., Hughes, Thomas P., and Pinch, Trevor J., eds. 1987a. *The Social Construction of Technological Systems*. Cambridge, MA: MIT Press.

Birch, Alan. 1963–1964. "Henry Bessemer and the Steel Revolution." *Nachrichten aus der Eisen-Bibliothek* (Schaffhausen) 28: 129–136; 30: 153–159.

Bloor, David C. 1976. *Knowledge and Social Imagery*. London: Routledge and Kegan Paul.

Bodewitz, Henk J. H. W., Buurma, H., and De Vries, Gerard H. 1987. "Regulatory Science and the Social Management of Trust in Medicine," in *The Social Construction of Technological Systems. New Directions in the Sociology and History of Technology*, W. E. Bijker, T. P. Hughes, and T. J. Pinch, eds. Cambridge, MA: MIT Press, 241–258.

Boltanski, Luc, and Thevenot, Laurent. 1987. *Les Economies de la Grandeur*. Paris: PUF, Cahiers du Centre d'Etudes de l'Emploi.

Boltanski, L., and Thevenot, L. 1991. *De la Justification les Economies de la Grandeur*. Paris: Gallimard.

Boorstin, Daniel J. 1973. *The Americans: The Democratic Experience*. New York: Random House.

Boucher, John Newton, ed. 1908. *A Century and a Half of Pittsburg and Her People*. Pittsburgh: Lewis.

Boucher, John Newton. 1924. *William Kelly: A True History of the So-called Bessemer Process*. Greensburg, PA: John Newton Boucher.

Boullier, D., Akrich, M., and Le Goaziou, V. 1990. *Représentation de l'utilisateur final et genèse des modes d'emploi*. Miméo, Ecole des Mines.

Bowker, Geof. 1987. "A Well-Ordered Reality: Aspects of the Development of *Schlumberger*, 1920–1939." *Social Studies of Science* 17: 611–655.

Bowker, Geof. 1988. 'Pictures from the Subsoil, 1939,' in "Picturing Power; Visual Depiction and Social Relations." *Sociological Review Monographs* 35, Gordon Fyfe and John Law, eds. 1986, 221–254.

Bowker, Geof. 1989. "L'Industrialisation de la Science," in *Elements d'Histoire des Sciences*, Michel Serres, ed. Paris: Bordas.

Boyer, P. 1985. *By the Bomb's early light. American thought and culture at the dawn of the Atomic Age*. New York: Pantheon.

Bright, A. A. 1949. *The Electrical Lamp Industry: Technological Change and Economic Development from 1800 to 1947*. New York: MacMillan.

Bright, A. A., and Maclaurin, W. R. 1943. "Economic Factors Influencing the Development and Introduction of the Fluorescent Lamp." *Journal of Political Economy* 51: 429–450.

Buchanan, Angus. 1991. "Theory and Narrative in the History of Technology," *Technology and Culture* 32: 365–376.

Butler, Samuel 1872 (paperback edition 1970). *Erewhon*. Harmondsworth: Penguin.

Callon, Michel. 1980. "Struggles and Negotiations to Define What is Problematic and What is Not: the Sociologic of Translation," in *The Social Process of Scientific Investigation*, vol. 4, K. Knorr, R. Krohn, and R. D. Whitley, eds. Dordrecht: Reidel, 197–219.

Callon, Michel. 1986a. "Some Elements of a Sociology of Translation: Domestication of the Scallops and the Fishermen of St. Brieuc Bay," in *Power, Action and Belief: a New Sociology of Knowledge?* J. Law, ed. London: Routledge and Kegan Paul, 196–233.

Callon, Michel. 1986b. "The Sociology of an Actor-Network: The Case of the Electric Vehicle," in *Mapping the Dynamics of Science and Technology: Sociology of Science in the Real World*, Michel Callon, John Law, and Arie Rip, eds. Basingstoke: Macmillan, 19–34.

Callon, Michel, and Latour, Bruno. 1981. "Unscrewing the Big Leviathan: How Actors Macrostructure Reality and How Sociologists Help Them To Do So," in *Advances in Social Theory and Methodology: Toward an Integration of Micro and Macro Sociologies*, K. Knorr-Cetina, and A. V. Cicourel, eds. London: Routledge and Kegan Paul, 277–303.

Callon, Michel, and Latour, Bruno. 1992. "Don't Throw Out the Baby with the Bath School: Reply to Collins and Yearley," in *Science as Practice and Culture*, A. Pickering, ed. Chicago: Chicago University Press.

Callon, Michel, and Law, John. 1989. "On the Construction of Sociotechnical Networks: Content and Context Revisited." *Knowledge and Society* 9: 57–83.

Callon, Michel, Law, John, and Rip, Arie, eds. 1986. *Mapping the Dynamics of Science and Technology*. Basingstoke: Macmillan.

Cambrosio, Alberto, Keating, Peter, and Mackenzie, Michael. (Forthcoming). "Scientific Practice in the Courtroom: The Construction of Sociotechnical Identities in a Biotechnology Patent Dispute."

Camp, J. M., and Francis, C. B. 1925. *The Making, Shaping and Treating of Steel*, 4th ed. Pittsburgh: Carnegie Steel Co.

Campbell, John. 1983. *Roy Jenkins, a Biography*. London: Weidenfeld and Nicolson.

Carlson, W. Bernard. 1983. "Edison in the Mountains: The Magnetic Ore Separation Venture, 1879–1900." *History of Technology* 8: 37–59.

Carlson, W. Bernard. 1988. "Thomas Edison as a Manager of R&D: The Case of the Alkaline Storage Battery, 1898–1915." *IEE Technology and Society* 7: 4–12.

Carlson, W. Bernard. 1991. *Innovation as a Social Process: Elihu Thomson and the Rise of the Electrical Industry, 1875–1900*. New York: Cambridge University Press.

Carlson, W. Bernard, and Gorman, Michael E. 1989. "Thinking and Doing at Menlo Park: Edison's Development of the Telephone, 1876–1878," in *Working at Inventing: Thomas A. Edison and the Menlo Park Experience*, W. S. Pretzer, ed. Dearborn, MI: Henry Ford Museum and Greenfield Village, 84–99.

Carlson, W. Bernard, and Gorman, Michael E. 1990. "Understanding Invention as a Cognitive Process: The Case of Thomas Edison and Early Motion Pictures, 1888–1891." *Social Studies of Science* 20: 387–430.

Cassady, Ralph. 1982. "Monopoly in Motion Picture Production and Distribution: 1908–1915," in *The American Movie Industry: The Business of Motion Pictures*, G. Kindem, ed. Carbondale, IL: Southern Illinois University Press, 25–67.

Chandler, Alfred D. 1977. *The Visible Hand: The Managerial Revolution in American Business*. Cambridge, MA: Harvard University Press.

Clapp, F. G. 1929. "Role and Structure in the Accumulation of Petroleum," in *Structure of Typical American Oil Fields: A Symposium on the Relation of Oil Accumulation to Structure, vol. 2*. London: Thomas Murby, 667–716.

Clarke, Basil. 1965. *Supersonic Flight*. London: Frederick Muller.

Clegg, Stewart R. 1989. *Frameworks of Power*. London: Newbury Park; New Delhi: Sage.

Clifford, James. 1983. "On Ethnographic Authority," *Representations* 1, 2.

Collins, H. M. 1981a. "The Place of the Core-Set in Modern Science: Social Contingency with Methodological Propriety in Science." *History of Science* 19: 6–19.

Collins, H. M. 1981b. "Stages in the Empirical Programme of Relativism," *Social Studies of Science* 11: 3–10.

Collins, H. M. 1985. *Changing Order*. Beverly Hills and London: Sage.

Collins, H. M., and Pinch, Trevor J. 1982. *Frames of Meaning: The Social Construction of Extraordinary Science*. Boston: Routledge and Kegan Paul.

Collins, H. M., and Yearley, Steven. 1992. "Epistemological Chicken," in *Science in Practice and Culture*, A Pickering, ed. Chicago: Chicago University Press.

Committee on Patents. 1942. *Hearings before the Committee on Patents United States Senate, 77th Congress, 2nd Session on S.2303 and S.2491, Part 9, 18–21 August 1942: 4753–5032*. Washington, D.C.: United States Government Printing Office.

Constant. Edward W. 1980. *The Origins of the Turbojet Revolution*. Baltimore, MD: Johns Hopkins University Press.

Constant, Edward W. 1983. "Scientific Theory and Technological Testability: Science, Dynamometers, and Water Turbines in the 19th Century." *Technology and Culture* 24: 183–198.

Crossman, Richard. 1975. *The Diaries of a Cabinet Minister; Vol. 1, Minister of Housing*. London: Hamish Hamilton and Jonathan Cape.

Culler, F. J., Mclain, S., eds. 1957. *Status Report on the Disposal of Radioactive Wastes*. Oak Ridge, TN: Oak Ridge National Laboratory, 57-3-114.

Culyer, A. J. 1985. "Editorial," *Nuffield/York Portfolios* 19. Nuffield Provincial Hospital Trust, London, p. 1.

Czitrom, Daniel J. 1982. *Media and the American Mind: From Morse to McLuhan*. Chapel Hill, NC: University of North Carolina Press.

Daumas, M. 1977. "Analyse historique de l'evolution des transports en commun dans la region parisienne de 1855 a 1939." Paris: Centre de documentation d'Histoire des Techniques.

Davis, Donald Finlay. 1988. *Conspicuous Production: Automobiles and Elites in Detroit, 1899–1933*. Philadelphia: Temple University Press.

Dean, John. 1979. "Controversy over Classification: A Case Study from the History of Botany," in *Natural Order: Historical Studies of Scientific Culture*, Barry Barnes and Steven Shapin, eds. London: Sage, 211–230.

"Dedication of Tablet Recalls Bessemer Patent Controversy." 1922. *Iron Trade Review* 71 (19 October): 1064.

Del Sesto, S. L. 1987. "Wasn't the Future of Nuclear Engineering Wonderful?" in *Imagining Tomorrow. History, Technology and the American Future*, J. J. Corn, ed. Cambridge, MA: MIT Press, 58–76.

Dennis, Michael Aaron. 1987. "Accounting for Research: New Histories of Corporate Laboratories and the Social History of American Science." *Social Studies of Science* 17: 479–518.

Dickson, W. K. L. 1933. "A Brief History of the Kinetograph, the Kinetoscope, and the Kineto-phonograph," in *A Technological History of Motion Pictures and Television*, R. Fielding, ed. Berkeley: University of California Press, 1980, 9–16. This article was originally published in *Journal of the Society of Motion Picture Engineers*, vol. 21 (Dec. 1933).

Divine, R. A. 1978. *Blowing on the wind. The nuclear test ban debate, 1954–1960*. New York: Oxford University Press.

Doctorow, E. L. 1985. *World's Fair*. New York: Ballantine Books.

Dorsett, J. D. 1950. "Insurance Problems with Atomic Energy Use," in *Industrial and Safety Problems of Nuclear Technology*, H. M. Shamos and S. G. Roth, eds. New York: Harper.

Dosi, Giovanni. 1982. "Technological Paradigms and Technological Trajectories." *Research Policy* 11: 147–162.

Douglas, Mary. 1973. *Natural Symbols: Explorations in Cosmology*. Harmondsworth: Penguin.

Douglas, Susan J. 1987. *Inventing American Broadcasting, 1899–1922*. Baltimore, MD: Johns Hopkins University Press.

Dredge, James. 1898. "Sir Henry Bessemer." *ASME Transactions* 19: 881–964.

Dutton, H. I. 1984. *The Patent System and Inventive Activity during the Industrial Revolution, 1750–1852*. Manchester: Manchester University Press.

Eisenstein, Elizabeth L. 1979. *The Printing Press as an Agent of Change: Communications and Culture in Early Modern Europe*. Cambridge: Cambridge University Press.

Eisinger, Peter K. 1988. *The Rise of the Entrepreneurial State: State and Local Economic Development Policy in the United States*. Madison, WI: University of Wisconsin Press.

Elias, Norbert. 1978. *The History of Manners*. Oxford: Blackwell.

Elzen, Boelie. 1986. "Two Centrifuges: A Comparative Study of the Social Construction of Artefacts." *Social Studies of Science* 16: 621–662.

Elzen, Boelie. 1988. *Scientists and Rotors: The Development of Biochemical Ultracentrifuges*. Ph.D. diss., University of Twente.

Engelhardt, H. Tristram, Jr., and Caplan, Arthur L., eds. 1987. *Scientific Controversies: Case Studies in the Resolution and Closure of Disputes in Science and Technology*. Cambridge: Cambridge Univesity Press.

Engelken, R. C. 1940. "Lighting the New York World's Fair", *Journal of Electrical Engineering* 59 (May 1940): 179–203.

"Fluctuations in the Prices of Crude and Finished Iron and Steel from January 1, 1898, to January 1, 1907." 1907. *Iron Age* supplement (10 January).

Foucault, Michel. 1975. *Surveiller et Punir, Naissance de la Prison*. Paris: Gallimard.

Fox, Richard W., and Lears, T. J. Jackson, eds. 1983. *The Culture of Consumption: Critical Essays in American History, 1880–1980*. New York: Pantheon.

Friedel, Robert, and Israel, Paul. 1986. *Edison's Electric Light: Biography of an Invention*. New Brunswick, NJ: Rutgers University Press.

Frontisi-Ducroux, F. 1975. *Dédale, Mythologie de l'artisan en Grèce Ancienne*. Paris: Maspéro-La Découverte.

Fujimura, Joan. 1987. "Constructing 'Do-able' Problems in Cancer Research: Articulating Alignment." *Social Studies of Science* 17: 257–293.

Galambos, Louis, and Pratt, Joseph. 1988. *The Rise of the Corporate Commonwealth: United States Business and Public Policy in the 20th Century*. New York: Basic Books.

Gallup, G. H. 1972. *The Gallup Polls: Public Opinion 1935–1971*. New York: Random House.

Gardner, Charles. 1981. *British Aircraft Corporation, a History*. London: Batsford.

Giddens, Anthony. 1984. *The Constitution of Society*. Cambridge: Polity Press.

Gilbert, G. Nigel, and Mulkay, Michael. 1984. *Opening Pandora's Box: A Sociological Analysis of Scientists' Discourse*. Cambridge: Cambridge University Press.

Gish, O. H. 1947 (1932). "Use of Geo-electric Methods in the Search for Oil," in *Early Geophysical Papers of the Society of Exploration Geophysics*. Tulsa, OK: Society of Exploration Geophysics, 497–508.

Glaser, Barney, and Strauss, Anselm. 1966. *Awareness of Dying*. Chicago: Aldine.

Gökalp, Iskender. 1992. "On the Analysis of Large Technical Systems." *Science, Technology & Human Values* 17 (winter): 57–78.

Gooding, David, Pinch, Trevor, and Schaffer, Simon, eds. 1989. *The Uses of Experiment: Studies in the Natural Sciences.* Cambridge: Cambridge University Press.

Gorman, Michael E., and Carlson, W. Bernard. 1990. "Interpreting Invention as a Cognitive Process: Alexander Graham Bell, Thomas Edison, and the Telephone, 1876–1878." *Science, Technology, and Human Values* 15 (spring 1990): 131–164.

Greiner, A. 1877. "Nomenclature of Steel." *Engineering and Mining Journal* 23 (3 March): 138–139.

Griffiths, R., et al. 1983. *NHS Management Inquiry.* London: DHSS.

Gunston, Bill. 1974. *Attack Aircraft of the West.* London: Ian Allen.

Hacker, B. C. 1987. *The Dragon's Tail. Radiation Safety in the Manhattan Project, 1942–1946.* Berkeley: University of California Press.

Harrison, W., and Hibben, S. G. 1938. "Efficient Tint Lighting With Fluorescent Tubes." *Electrical World* 110 (May 1938): 1523–1530.

Hastings, Stephen. 1966. *The Murder of TSR 2.* London: Macdonald.

Heiland, C. A. 1940. *Geophysical Exploration.* New York: Hafner.

Hellrigel, Mary Ann. 1989. "Creating an Industry: Thomas A. Edison and His Electric Light System." Master's thesis, University of California, Santa Barbara.

Hendricks, Gordon. 1966. *The Kinetoscope: America's First Commercially Successful Motion Picture Exhibitor.* New York: The Beginnings of the American Film.

Hewish, John. 1987. "From Cromford to Chancery Lane: New Light on the Arkwright Patent Trials." *Technology and Culture* 28: 80–86.

Hewlett, R. G. 1978. *Federal Policy for the disposal of highly radioactive wastes from commercial nuclear power plants.* Washington, DC: Department of Energy, DOE/MA-0153.

Hewlett, R. G., and Duncan, F. 1969. *Atomic Shield, 1947–1952.* University Park and London: Pennsylvania State University Press.

Hindle, Brooke. 1981. *Emulation and Invention.* New York: New York University Press.

Hine, T. 1986. *Populuxe.* New York: Knopf.

Hollander, W. 1981. *Abel Wolman. His Life and Philosophy, an Oral History.* Chapel Hill, NC: Universal Printing and Publishing Company.

Holley, Alexander L. 1865. *A Treatise on Ordnance and Armor.* New York: D. Van Nostrand.

Holley, Alexander L. 1868. *The Bessemer Process and Works in the United States.* New York: D. Van Nostrand.

Holley, Alexander L. 1872. "Bessemer Machinery." *Journal of the Franklin Institute* 94: 252–265, 391–399; (1873): 233–241.

Holley, Alexander L. 1873a. *Bessemer Machinery.* Philadelphia: Merrihew.

Holley, Alexander L. 1873b. "Tests of Steel." *American Institute of Mining Engineers Transactions* 2: 116–122.

Holley, Alexander L. 1875. "What is Steel?" *American Institute of Mining Engineeers Transactions* 4: 138–149.

Holton, Gerald. 1978. "Subelectrons, Presuppositions, and the Millikan-Ehrenhaft Dispute." *Historical Studies in the Physical Sciences* 9: 161–224.

Hounshell, David A., and Smith, John Kenly, Jr. 1988. *Science and Corporate Strategy: Du Pont R&D, 1902–1980.* Cambridge: Cambridge University Press.

Howe, Henry M. 1875. "What is Steel?" *Engineering and Mining Journal* 20 (28 August; 4, 11, 18 September 1875): 213, 235–236, 258–259, 282–283.

Howe, Henry M. 1876. "The Nomenclature of Iron." *American Institute of Mining Engineers Transactions* 5: 515–537.

Howe, Henry M. 1891. *The Metallurgy of Steel,* 2nd edition, revised. New York: Scientific Publishing.

Hughes, Thomas P. 1971. *Elmer Sperry: Inventor and Engineer.* Baltimore, MD: Johns Hopkins University Press.

Hughes, Thomas P. 1977. "Edison's Method," in *Technology at the Turning Point,* W. B. Pickett, ed. San Francisco: San Francisco Press, 5–22.

Hughes, Thomas P. 1983. *Networks of Power: Electrification in Western Society, 1880–1930.* Baltimore, MD: Johns Hopkins University Press.

Hughes, Thomas P. 1986a. "The Seamless Web: Technology, Science, Etcetera, Etcetera." *Social Studies of Science* 16: 281–292.

Hughes, Thomas P. 1986b. "Machines and Medicine. A Projection of Analogies Between Electric Power Systems and Health Care Systems." *International Journal of Technology Assessment in Health Care* 2: 285–295.

Hughes, Thomas P. 1987. "The Evolution of Large Technological Systems," in Bijker, et al., eds. 1987a: 51–82.

Hughes, Thomas P. 1989. *American Genesis: A Century of Invention and Technological Enthusiasm, 1870–1970.* New York: Viking Penguin.

Hull, David L. 1988. *Science as a Process: An Evolutionary Account of the Social and Conceptual Development of Science.* Chicago: University of Chicago Press.

Hung, Robert W. 1876. "A History of the Bessemer Manufacture in America." *American Institute of Mining Engineers Transactions* 5: 201–216.

Ignatieff, Michael. 1978. *A Just Measure of Pain: The Penitentiary in the Industrial Revolution, 1750–1850.* New York: Pantheon.

Inman, G. E., and Thayer, R. N. 1938. "Low-Voltage Fluorescent Lamps." *Journal for Electrical Engineering* 57 (June 1938): 245–248.

"The Invention of the Bessemer Process." *Engineering* 6 (27 March 1896): 413–414.

Jacobs, Dany. 1988. *Gereguleerd Staal: Nationale en Internationale Economische Regulering in de Westeuropese Staalindustrie, 1750–1950.* Ph.D. diss., University of Nijmegen.

Jacobs, Lewis. 1968 (1939). *The Rise of the American Film: A Critical History.* New York: Teachers College Press, 1939; reprinted 1968.

Jeans, W. T. 1884. *The Creators of the Age of Steel.* New York: C. Scribner's Sons; London: Chapman and Hall.

Jehl, Francis. 1937. *Menlo Park Reminscences*, 3 vols. Dearborn, MI: Edison Institute, 1937–1941.

Jenkins, Reese V. 1984. "Elements of Style: Continuities in Edison's Thinking." *Annals of the New York Academy of Sciences* 424: 149–162.

Jenkins, Reese V., and Israel, Paul. 1948. "Thomas A. Edison: Flamboyant Inventor." *IEEE Spectrum*, December, 74–79.

Jenkins, Reese V., et al. 1989. *The Papers of Thomas A. Edison. Vol. 1: The Making of an Inventor, February 1847–June 1873*. Baltimore, MD: Johns Hopkins University Press.

Johnson, Terence J. 1972. *Professions and Power*. London: Macmillan.

Johnson, Terence J. 1977. "The Professions in the Class Structure," in *Industrial Society: Class, Cleavage, and Control*, Richard Scase, ed. New York: St. Martin's, 93–110.

Jones, Edgar. 1988. "The Transition from Wrought Iron to Steel Technology at the Dowlais Iron Company, 1850–1890," in *The Challenge of New Technology: Innovation in British Business Since 1850*, Jonathan Liebenau, ed. Gower: Aldershot, 43–57.

Josephson, Matthew. 1959. *Edison: A Biography*. New York: McGraw-Hill.

Kasson, John. 1978. *Amusing the Million: Coney Island at the Turn of the Century*. New York: Hill and Wang.

Kindem, G., ed. 1982. *The American Movie Industry: The Business of Motion Pictures*. Carbondale, IL: Southern Illinois University Press.

Kopp, C. 1979. "The origins of the American scientific debate over fallout." *Social Studies of Science* 9: 403–422.

Kraus, S., Mehling, R., and El-Assal, E. 1963. "Mass Media and the Fallout Controversy." *Public Opinion Quarterly* 27: 191–206.

Kuklick, Henrika. 1983. "The Sociology of Knowledge: Retrospect and Prospect." *Annual Review of Sociology* 9: 287–310.

Lamb, Robert K. 1952. "The Entrepreneur and the Community," in *Men in Business: Essays in the History of Entrepreneurship*, W. Miller, ed. Cambridge, MA: Harvard University Press, 91–119.

Lange, Ernest F. 1913. "Bessemer, Göransson and Mushet: A Contribution to Technical History." *Memoirs Manchester Lit & Phil Society* 57, no. 17: 1–44.

Latour, Bruno. 1987. *Science in Action. How to follow scientists and engineers through society*. Milton Keynes: Open University Press; and Cambridge, MA: Harvard University Press.

Latour, Bruno. 1988a. "How to Write *The Prince* for Machines as Well as for Machinations," in *Technology and Social Change*, Brian Elliot, ed. Edinburgh: Edinburgh University Press.

Latour, Bruno. 1988b. "A Relativist Account of Einstein's Relativity." *Social Studies of Science* 18: 3–45.

Latour, Bruno. 1988c. *Irreductions*, published with *The Pasteurization of France*, translated by A. Sheridan and J. Law. Cambridge, MA: Harvard University Press.

Latour, Bruno. 1990. "Drawing Things Together," in *Representation in Scientific Practice*. Michael Lynch and Steve Woolgar, eds. Cambridge, MA: MIT Press, 19–68.

Latour, Bruno. 1992. *Aramis ou l'amour des techniques*. Paris: La Découvertè.

Latour, Bruno, Mauguin, P., and Teil, Genvieve 1992. "A Note on Socio-Technical Graphs." *Social Studies of Science* 22: 33–57.

Law, John. 1987a. "On the Social Explanation of Technical Change: The Case of the Portuguese Maritime Expansion." *Technology and Culture* 28: 227–252.

Law, John. 1987b. "Technology and Heterogeneous Engineering: The Case of Portuguese Expansion," in *The Social Construction of Technological Systems*, W. E. Bijker, T. P. Hughes, and T. J. Pinch, eds. Cambridge, MA: MIT Press, 111–134.

Law, J. 1988. "The Anatomy of a Sociotechnical Struggle: The Design of the TSR2," in *Technology and Social Process*, B. Elliott, ed. Edinburgh: Edinburgh University Press, 44–69.

Law, John. 1991a. "Theory and Narrative in the History of Technology: Response," *Technology and Culture* 32: 377–384.

Law, John. 1991b. "Power, Discretion and Strategy: Management Discourse in a Formal Organisation," in *Power, Technology and the Modern World*, J. Law, ed. London: Routledge.

Law, John. 1992. "The Olympus 320 Engine: A Case Study in Design, Autonomy and Organizational Control," in *Technology and Culture*.

Leroi-Gourhan, A. 1964. *Le Geste et la Parole*. Paris: Albin-Michel.

Lord, W. M. 1945–1947. "The Development of the Bessemer Process in Lancashire, 1856–1900." *Newcomen Society Transactions* 25: 163–180.

Lundvall, Bengt-Åke. 1988. "Innovation as an Interactive Process: From User-Producer Interaction to the National System of Innovation," in *Technical Change and Economic Theory*, Giovanni Dosi, et al., eds. London: Pinter, 349–369.

Lyell, Charles. *Principles of Geology*, 3 vols., London, 1830–1832.

Lynch, Michael, and Woolgar, Steve eds. 1990. *Representation in Scientific Practice*. Cambridge, MA: MIT Press.

Mack, P. 1990. *Viewing the Earth. The Social Construction of the Landsat Satellite System*. Cambridge, MA: MIT Press.

MacKenzie, Donald. 1978. "Statistical Theory and Social Interests: A Case Study." *Social Studies of Science* 8: 35–83.

MacKenzie, Donald. 1987. "Missile Accuracy: A Case Study in the Social Processes of Technological Change," in *The Social Construction of Technological Systems: New Directions in the Sociology and History of Technology*, W. E. Bijker, T. P. Hughes, and T. J. Pinch, eds. Cambridge, MA: MIT Press, 195–222.

MacKenzie, Donald. 1989. "From Kwajalein to Armageddon? Testing and the social construction of missile accuracy," in *The Uses of Experiment*, D. Gooding, T. J. Pinch, and S. Schaffer, eds. Cambridge: Cambridge University Press.

MacKenzie, Donald. 1990a. *Inventing Accuracy: A Historical Sociology of Nuclear Missile Guidance*. Cambridge, MA: MIT Press.

MacKenzie, Donald. 1990b. "Economic and Sociological Explanation of Technical Change." Paper presented to Meeting on Firm Strategy and Technical Change: Micro Economics or Micro Sociology? at Manchester, 27–27 September.

MacKenzie, Donald. 1990c. "Negotiating Arithmetic, Deconstructing Proof: The Sociology of Mathematics and Information Technology." Mimeo, Edinburgh University.

MacKenzie, Donald, and Barnes, Barry. 1979. Scientific Judgement: The Biometry-Mendelism Controversy," in *Natural Order: Historical Studies of Scientific Culture*, Barry Barnes and Steven Shapin, eds. London: Sage, 191–210.

MacKenzie, Donald, and Spinardi, Graham. 1988. "The Shaping of Nuclear Weapon System Technology: US Fleet Ballistic Missile Guidance and Navigation." *Social Studies of Science* 18: 419–463; 581–624.

MacKenzie, Donald, and Wajcman, Judy, eds. 1985. *The Social Shaping of Technology a Reader*. Milton Keynes: Open University Press.

Macleod, Christine. 1988. *Inventing the Industrial Revolution: The English Patent System, 1660–1800*. Cambridge: Cambridge University Press.

Maier, Charles S. 1975. *Recasting Bourgeois Europe: Stabilization in France, Germany, and Italy in the Decade after World War I*. Princeton: Princeton University Press.

Maier, Charles S. 1987. *In Search of Stability: Explorations in Historical Political Economy*. New York: Cambridge University Press.

Mann, Michael. 1986. *The Sources of Social Power. Vol. 1. A History of Power from the Beginning to A.D. 1760*. Cambridge: Cambridge University Press.

May, Lary. 1980. *Screening out the Past: The Birth of Mass Culture and the Motion Picture Industry*. New York: Oxford University Press.

Mayntz, Renate, and Hughes, Thomas P., eds. 1988. *The Development of Large Technical Systems*. Frankfurt: Campus Verlag.

Mazuzan, G. T., and Walker, J. S. 1984. *Controlling the Atom. The Beginnings of Nuclear Regulation, 1946–1962*. Berkeley: University of California Press.

McCracken, Grant. 1988. *Culture and Consumption: New Approaches to the Symbolic Character of Consumer Goods and Activities*. Bloomington: Indiana University Press.

McGaw, Judith A. 1985. "Accounting for Innovation: Technological Change and Business Practice in the Berkshire County Paper Industry." *Technology and Culture* 26: 703–725.

McHugh, Jeanne. 1980. *Alexander Holley and the Makers of Steel*. Baltimore, MD: Johns Hopkins University Press.

Metcalf, William. 1876. "Can the Commercial Nomenclature of Iron Be Reconciled to the Scientific Definitions of the Terms Used to Distinguish the Various Classes?" *American Institute of Mining Engineers Transactions* 5: 355–365.

Metcalf, William, et al. 1880. "Discussion on Steel Rails." *American Institute of Mining Engineers Transactions* 9: 529–608.

Metlay, D. 1985. "Radioactive waste management policy making," in *Managing the National's Commercial High-Level Radioactive Waste*. Washington, DC: Office of Technology Assessment (OTA), 199–244.

Millard, A. J. 1990. *Edison and the Business of Innovation.* Baltimore: Johns Hopkins University Press.

Misa, Thomas J. 1985. "Military Needs, Commercial Realities, and the Development of the Transistor, 1948–1958," in *Military Enterprise and Technological Change,* Merritt Roe Smith, ed. Cambridge, MA: MIT Press, 253–287.

Misa, Thomas J. 1987. *Science, Technology and Industrial Structure: Steelmaking in America, 1870–1925.* Ph.D. diss., University of Pennsylvania.

Misa, Thomas J. 1988a. "How Machines Make History, and How Historians (and Others) Help Them to Do So." *Science, Technology and Human Values* 13: 308–331.

Misa, Thomas J. 1988b. "The Construction and Destruction of a Heterogeneous Network: The Case of High Speed Tool Steel." Paper presented to European Association for the Study of Science and Technology, Amsterdam, 16–19 November.

Misa, Thomas J. 1992. "Theories of Technological Change: Parameters and Purposes." *Science, Technology, and Human Values* 17 (winter): 3–12.

Moon, P. 1936 (revised ed. 1961). *The Scientific Basis of Illuminating Engineering.* New York: Dover Publications.

Morrell, Jack, and Thackray, Arnold. 1981. *Gentlemen of Science: Early Years of the British Association for the Advancement of Science.* Oxford: Clarendon Press.

Mulkay, Michael. 1979. "Knowledge and Utility: Implications for the Sociology of Knowledge," *Social Studies of Science* 9: 63–80.

Mulkay, Michael. 1985. *The Word and the World.* London: George Allen and Unwin.

Musser, Charles. 1991. *Before the Nickelodeon: Edwin S. Porter and the Edison Manufacturing Company.* Berkeley: University of California Press.

Mulkay, Michael, Pinch, Trevor, and Ashmore, Malcolm. 1987. "Colonizing the Mind. Dilemmas in the Application of Social Science." *Social Studies of Science* 17: 231–256.

Nelson, Richard R, and Winter, Sidney G. 1982. *An Evolutionary Theory of Economic Change.* Cambridge, MA: Belknap/Harvard University Press.

Noble, David F. 1977. *America by Design: Science, Technology and the Rise of Corporate Capitalism.* New York: Knopf.

Noble, David. 1984. *Forces of Production: A Social History of Industrial Automation.* New York: Knopf.

Norman, David. 1988. *The Psychology of Everyday Things.* New York: Basic Books.

Pearse, John B. 1872. "The Manufacture of Iron and Steel Rails." *American Institute of Mining Engineers Transactions* 1: 162–169.

Peiss, Kathy Lee. 1986. *Cheap Amusements: Working Women and Leisure in Turn-of-the-Century New York.* Philadelphia: Temple University Press.

Perrow, Charles. 1984. *Normal Accidents: Living with High-Risk Technologies.* New York: Basic Books.

Pfau, R. 1984. *No Sacrifice Too Great: The Life of Lewis L. Strauss.* Charlottesville, VA: University Press of Virginia.

Philip, Cynthia Owen. 1985. *Robert Fulton: A Biography.* New York: Franklin Watts.

Pinch, Trevor J. 1986. *Confronting Nature: The Sociology of Solar-Neutrino Detection.* Dordrecht: D. Reidel.

Pinch, Trevor, Ashmore, Malcolm, and Mulkay, Michael. 1987. "Social Technologies: To Test or Not to Test, That Is the Question." Presented to the International Workshop on the Integration of Social and Historical Studies of Technology, University of Twente, 3–5 September.

Pinch, Trevor J., and Bijker, W. E. 1984. "The Social Construction of Facts and Artefacts: Or How the Sociology of Science and the Sociology of Technology Might Benefit Each Other." *Social Studies of Science.* 14: 399–441.

Pinch, Trevor J., and Bijker, Wiebe E. 1986. "Science, Relativism and the New Sociology of Technology." *Social Studies of Science* 16: 347–360.

Pinch, Trevor J., and Bijker, Wiebe E. 1987. "The Social Construction of Facts and Artifacts: Or How the Sociology of Science and the Sociology of Technology Might Benefit One Another," in *The Social Construction of Technological Systems*, Wiebe E. Bijker, Thomas P. Hughes, and Trevor J. Pinch, eds. Cambridge, MA: MIT Press, 17–50.

Pool, Ithiel de Sola. 1983. *Technologies of Freedom.* Cambridge, MA: Belknap/ Harvard University Press.

Post, Robert C. 1976. *Physics, Patents, and Politics: A Biography of Charles Grafton Page.* New York: Science History.

Pratt, Wallace E. 1940. "Geology in the Petroleum Industry." *Bulletin of the American Association of Petroleum Geologists* 24: 1209–1240.

Price Waterhouse and Comshare. 1986. "Management Budgeting for General Managers." Available from Price Waterhouse, London.

Prime, Frederick, Jr. 1875. "What Steel Is." *American Institute of Mining Engineers Transactions* 4: 328–339.

Rae, John B. 1965. *The American Automobile: A Brief History.* Chicago: University of Chicago Press.

Ramsaye, Terry. 1926. *A Million and One Nights: A History of the Motion Picture.* New York: Simon and Schuster.

Reed, Bruce, and Williams, Geoffrey. 1971. *Denis Healey and the Policies of Power.* London: Sidgewick and Jackson.

Reich, L. 1985. *The Making of American Industrial Research: Science and Business at G.E. and Bell, 1876–1926.* Cambridge: Cambridge University Press.

Ridenour, L. N. 1950. "How effective are radioactive poisons in warfare?" *Bulletin of Atomic Scientists* 6: 199–202, 224.

Robertson, James Oliver. 1980. *American Myth, American Reality.* New York: Hill & Wang.

Rogers, R. P. 1980. *The Development and Structure of the U.S. Electric Lamp Industry, Bureau of Economics Staff Report of the Federal Trade Commission.* Washington, D.C.: U.S. Government Printing Office.

Rosenberg, Nathan. 1982. *Inside the Black Box: Technology and Economics.* Cambridge: Cambridge University Press.

Rosenzweig, Roy. 1983. *Eight Hours for What We Will: Workers and Leisure in an Industrial City, 1870–1920*. New York: Cambridge University Press.

Rosi, E. J. 1965. "Mass and Attentive Opinion on Nuclear Weapon Tests and Fallout, 1954–1963." *Public Opinion Quarterly* 29: 280–298.

Rudwick, Martin J. S. 1985. *The Great Devonian Controversy: The Shaping of Scientific Knowledge among Gentlemanly Specialists*. Chicago: University of Chicago Press.

Russell, Conrad, ed. 1973. *The Origins of the English Civil War*. London: Macmillan.

Russell, Stewart. 1986. "The Social Construction of Artefacts: a Response to Pinch and Bijker." *Social Studies of Science* 16: 331–346.

Sahal, D. 1981. *Patterns of Technological Innovation*. Reading, MA: Addison-Wesley.

Sandberg, C. P. 1880. "Rail Specifications and Rail Inspection in Europe." *American Institute of Mining Engineers Transactions* 9: 193–248.

Sandberg, C. P., et al. 1881. "Iron and Steel Considered as Structural Materials— A Discussion." *American Institute of Mining Engineers Transactions* 10: 361–411.

Schifrin, Art. 1983a. "Researching and Restoring Pioneering Talking Pictures: The 70th Anniversary of the Theatrical Release of the Kinetophone." *Journal of the Society of Motion Picture and Television Engineers* 92: 739–751.

Schifrin, Art. 1983b. "The Trouble with the Kinetophone." *American Cinematographer* 64 (September): 50–54, 115.

Scranton, Philip. 1991. "Theory and Narrative in the History of Technology: Comment." *Technology and Culture* 32: 385–393.

Secord, James A. 1986. *Controversy in Victorian Geology: The Cambrian-Silurian Dispute*. Princeton, NJ: Princeton University Press.

Segerstråle, Ullica. 1986. "Colleagues in Conflict: An 'In Vivo' Analysis of the Sociobiology Controversy." *Biology and Philosophy* 1: 53–87.

Shapin, S., and Schaffer, S. 1985. *Leviathan and the Air Pump: Hobbes, Boyle and the Experimental Life*. Princeton: Princeton University Press.

Shelley, Mary. 1816 (pocket edition, 1983). *Frankenstein*. Harmondsworth: Penguin.

Shrum, Wesley, and Morris, Joan. 1990. "Organizational Constructs for the Assembly of Technological Knowledge," in *Theories of Science in Society*, Susan E. Cozzens and Thomas F. Gieryn, eds. Bloomington and Indianapolis: Indiana University Press, 235–257.

Siemens, C. W. 1868. "The Regenerative Gas Furnace as Applied to the Manufacture of Cast Steel." *Journal of the Chemical Society* (London) n.s. 6: 279–308.

Sigaut, F. 1984. "Essai d'identification des instrument a bras au travail de sol." *Cahiers ORSTOM, Science Humaines* 22, 3/4: 359–374.

Simon, H. A. 1969. *The Sciences of the Artificial*. Cambridge, MA: MIT Press.

Singer, Ben. 1988. "Early Home Cinema and the Edison Home Projecting Kinetoscope." *Film History* 2: 37–69.

Sklar, Robert. 1975. *Movie-Made America: A Social History of American Movies*. New York: Random House.

Slide, Anthony. 1970. *Early American Cinema*. New York: A. S. Barnes.

Smith, Dorothy. 1974. "The Social Construction of Documentary Reality." *Sociological Enquiry* 44(4): 257–268.

Star, Susan Leigh. 1989. *Regions of the Mind: Brain Research and the Quest for Scientific Certainty*. Stanford: Stanford University Press.

Staudenmaier sj, John. 1985. *Technology's Storytellers: Reweaving the Human Fabric*. Cambridge, MA: MIT Press.

Stone, L. 1972. *The Causes of the English Revolution, 1529–1642*. New York: Harper.

Stone, L. 1973. *Family and Fortune: studies in aristocratic finance in the sixteenth and seventeenth centuries*. Oxford: Clarendon.

Stoughton, Bradley. 1908. *The Metallurgy of Iron and Steel*. New York: Hill.

Stoughton, Bradley. 1934. *The Metallurgy of Iron and Steel*. 4th edition. New York: McGraw-Hill.

Strauss, Anselm. 1978. *Negotiations*. San Francisco: Jossey-Bass.

Suchman, Lucy. 1987. *Plans and Situated Actions. The Problem of Human Machine Communication*. Cambridge: Cambridge University Press.

Supply of Military Aircraft, 1955. Cmd. 9388. London: Her Majesty's Stationery Office.

Susman, Warren I. 1984. *Culture as History: The Transformation of American Society in the Twentieth Century*. New York: Pantheon.

Swann, John. 1887. *An Investor's Notes on American Railroads*. New York: Putnam.

Swank, James M. 1964 (1982). *History of the Manufacture of Iron in All Ages*. New York: Burt Franklin.

Tate, Alfred O. 1938. *Edison's Open Door*. New York: E. P. Dutton.

Tawney, Richard Henry. 1960. *Religion and the rise of capitalism: An historical study*. London: Murray.

Temin, Peter. 1964. *Iron and Steel in Nineteenth-Century America: An Economic Inquiry*. Cambridge, MA: MIT Press.

Thomas, Donald E. 1987. *Diesel: technology and society in industrial Germany*. Tuscaloosa, AL: University of Alabama Press.

Titus, C. A. 1986. *Bombs in the Backyard. Atomic Testing and American Politics*. Reno and Las Vegas, NV: University of Nevada Press.

Todd, Edmund N. 1987. "A Tale of Three Cites: Electrification and the Structure of Choice in the Ruhr, 1886–1900." *Social Studies of Science* 17: 387–412.

Todd, Edmund N. 1989. "Industry, State, and Electrical Technology in the Ruhr circa 1900." *Osiris* 5: 243–259.

Toure, A. 1985. *Les petits metiers d'Abidjan*. Paris: Edition Karthala.

Trevor-Roper, Hugh R. 1951. "The Elizabethan Aristocracy: an anatomy anatomized." *Economic History Review*, 2nd series, 3: 279–298.

Trevor-Roper, Hugh R. 1953. "The Gentry, 1540–1640." *Economic History Review*, supplement 1.

Tweedale, Geoffrey. 1984. "Sir Henry Bessemer." *Dictionary of Business Biography.* London: Butterworths, I: 309–314.

Tweedale, Geoffrey. 1987. *Sheffield Steel and America: A Century of Commercial and Technological Interdependence, 1830–1930.* Cambridge: Cambridge University Press.

U.S. Atomic Energy Commission (USAEC). 1949a. *Reporting of the handling of radioactive waste materials in the United States Atomic Energy Program.* Report AEC 180-1, October 17.

USAEC. 1949b. *Reporting of the handling of radioactive waste materials in the United States Atomic Energy Program.* Report AEC 180-2, October 14.

USAEC. 1949c. *Handling Radioactive Wastes in the Atomic Energy Program.* Washington DC: Government Printing Office (GPO).

USAEC. 1956a. *Disposal of Radioactive Waste.* February 3.

USAEC. 1956b. *Disposal of Radioactive Waste.* Report AEC 180-5, March 30.

USAEC. 1956c. *Disposal of Radioactive Wastes in the US Atomic Energy Program.* (WASH-408). Washington, DC: Government Printing Office, May 17.

USAEC. 1957a. *Atomic Energy Commission Handling and Disposal of Radioactive Wastes.* Report AEC 180-6, June 14.

USAEC. 1957b. *Status Report on Handling and Disposal of Radioactive Wastes in the AEC Program.* (WASH-742). Washington, DC: Government Printing Office, August.

USAEC. 1958. *First Meeting of the AEC Waste Disposal Working Group (WDWG).* Report AEC 719-20, April 3.

USAEC. 1960a. *Annual Report to Congress of the Atomic Energy Commission for 1959.* Washington, DC: Government Printing Office.

USAEC. 1960b. *Opinion and final decision in the matter of industrial waste disposal corporation.* Report AEC-R, 42-23, June 27.

USAEC. 1960c. *Letter to Committee on Waste Disposal, NAS-NRC, regarding land disposal of radioactive wastes.* Report AEC 180-13, September 20.

USAEC. 1960d. *Land Disposal of Radioactive Waste—Addendum to AEC 180-3.* Report AEC 180-14, October 27.

U.S. Department of Energy (DOE) Archives. *Energy History Collection.*

U.S. DOE Archives. *Records of the US Atomic Energy Commission,* Record group 326, Collection 'Secretariat' (SECY)—Materials 12 (Waste Processing and Disposal).

U.S. Joint Committee on Atomic Energy (JCAE). 1959. *Hearings on Industrial Radioactive Waste Disposal.* 86th Congress, first session, January 28–30, and February 2–3. Washington, DC: Government Printing Office (GPO).

U.S. National Academy of Sciences (NAS) Archives. *Records of the committee on (geologic) waste disposal.*

USNAS Archives. *Records of the committee on disposal and dispersal of radioactive wastes.*

USNAS. 1956a. *The Biological Effects of Atomic Radiation: Summary Reports.* Washington DC: National Academy of Sciences.

USNAS. 1956b. *The Biological Effects of Atomic Radiation: A Report to the Public.* Washington DC: National Academy of Sciences.

USNAS. 1957a. *The Disposal of Radioactive Waste on Land.* Washington DC: National Academy of Sciences.

USNAS. 1957b. "Proceedings of the Princeton Conference on Disposal of Radioactive Waste Products, September 10–12, 1955, Princeton University, New Jersey," in *The Disposal of Radioactive Waste on Land.* Washington, DC: National Academy of Sciences.

USNAS. 1957c. *Status Report on the Disposal of Radioactive Wastes.* Washington, DC: National Academy of Sciences.

U.S. National Archives. *Records of the US Joint Committee on Atomic Energy.* Record group 128.

U.S. National Archives. Records of the US atomic Energy Commission. Record group 326.

Usselman, Steven W. 1985. *Running the Machine: The Management of Technological Innovation on American Railroads, 1860–1910.* Ph.D. diss., University of Delaware.

Vergragt, Philip J. 1988. "The Social Shaping of Industrial Innovations." *Social Studies of Science* 18: 483–513.

Wachhorst, Wyn. 1981. *Thomas Alva Edison: An American Myth.* Cambridge, MA: MIT Press.

Weart, S. R. 1988. *Nuclear Fear. History of Images.* Cambridge, MA: Harvard University Press.

Wertime, Theodore A. 1961. *The Coming of the Age of Steel.* Leiden: E. J. Brill.

West, P. 1986. "Clinical Budgeting—A Critique." Paper presented to the Health Economists' Study Group, University of Bath, July.

Wickings, I. 1983. "Griffiths Report: Consultants Face the Figures." *Health and Social Service Journal,* December 8.

Wickings, I., Childs, T., Coles, J., and Wheatcroft, C. 1985. *Experiments Using Pacts in Southend and Oldham HAs,* CASPE Research Paper, CASPE, King Edward's Hospital Trust Fund, London.

Wickings, I., and Coles, J. 1985. "The Ethical Imperative of Clinical Budgeting." *Nuffield/York Portfolios* 10: 1–8. Nuffield Provincial Hospital Trusts, London.

Wiebe, Robert. 1967. *The Search for Order, 1877–1920.* New York: Hill and Wang.

Wilford, J. N. 1981. *The Mapmakers.* New York: Knopf.

Williams, Geoffrey, Gregory, Frank, and Simpson, John. 1969. *Crisis in Procurement: A Case Study of the TSR-2.* London:Royal United Service Institution.

Williams, Rosalind H. 1982. *Dream Worlds: Mass Consumption in Late Nineteenth-Century France.* Berkeley: University of California Press.

Williams, W. Mattieu. 1890. *The Chemistry of Iron and Steel Making.* London: Chatto and Windus.

Wilson, Harold. 1971. *The Labour Government, 1964–1970: A Personal Record.* London: Weidenfeld and Nicolson and Michael Joseph.

Winch, P. 1958. *The Idea of a Social Science and Its Relation to Philosophy.* London: Routledge and Kegan Paul.

Winner, Langdon. 1977. *Autonomous Technology*. Cambridge, MA: MIT Press.

Winner, Langdon. 1980. "Do Artefacts Have Politics?" *Daedalus* 109: 121–136.

Wittgenstein, L. 1974. *Philosophical Investigations*, 3rd edition. Oxford: Basil Blackwell.

Wood, Derek. 1975. *Project Cancelled*. London: Macdonald's and Janes.

Woolgar, Steve. 1983. "Irony in the Social Study of Science," in *Science Observed: Perspectives on the Social Study of Science*, K. D. Knorr and M. Mulkay, eds. Beverly Hills and London: Sage, 239–266.

Woolgar, Steve, ed. 1988. *Knowledge and Reflexivity: New Frontiers in the Sociology of Knowledge*. Beverly Hills and London: Sage.

Contributors

Madeleine Akrich is a research fellow at the Centre de Sociologie de l'Innovation of the Ecole Normale Supérieure des Mines in Paris. Her research is on the sociology of technology; she has published several articles on the process of technology transfer to less-developed countries and on user representations during the innovation process.

Malcolm Ashmore is lecturer in sociology at Loughborough University of Technology. His interests are in the sociology of scientific knowledge, expertise, and reflexivity; currently, he is investigating the role of the debunker in cases of discovered "fraud." He has published *The Reflexive Thesis* (University of Chicago Press, 1989) and, with Michael Mulkay and Trevor Pinch, *Health and Efficiency: A Sociology of Health Economics* (Open University Press, 1989).

Wiebe E. Bijker is associate professor at the University of Limburg, The Netherlands. His present research focuses on the implications of recent work in the sociology and history of technology for issues of control, intervention, and social change; his work encompasses technology assessment, ethics, and theories of society.

Geof Bowker is director of the National Archive for the History of Computing at the Center for the History of Science, Technology and Medicine, University of Manchester. His research interests include the history of cybernetics and computing. He has written extensively on the history of Schlumberger.

Michel Callon is professor of sociology at the Ecole Normale Supérieure des Mines in Paris and director of the Centre de Sociologie de l'Innovation. His current research focuses on the dynamics of techno-economic networks and on scientific research policies. He has edited, with John Law and Arie Rip, *Mapping the Dynamics of Science and Technology* (Macmillan, 1986) and *La Science et ses réseaux* (La Découverte, 1989).

W. Bernard Carlson is assistant professor of humanities in the School of Engineering and Applied Science at the University of Virginia. His area of expertise is the history of American technology and business, and much of his research has focused on understanding the social and cognitive dimensions of invention and innovation. He is author of *Innovation as a Social Process: Elihu Thomson and the Rise of General Electric, 1870–1900* (Cambridge University Press, 1991).

Adri de la Bruhèze is currently a guest researcher at the Twente University of Technology in The Netherlands. He is interested in the history, sociology, and politics of technology, and his current research is on the history of radioactive waste technology in the United States and The Netherlands.

Bruno Latour was trained in philosophy and anthropology and is now professor at the Ecole Normale Supérieure des Mines in Paris and at the Science Studies Program of the University of California, San Diego. He is the author of *Laboratory Life* (1979, with Steve Woolgar), *Science in Action* (1987), and *The Pasteurization of France* (1988). He is currently working on technology and on the social history of French science.

John Law is professor of sociology and director of the Centre for Technological and Organisational Practice at the University of Keele. His current research interests include the management and organization of scientific research, the history of military aviation, and the relationship between power and technology. His most recent publication is an edited collection on the last topic called *A Sociology of Monsters: Essays on Power, Technology and Domination* (Routledge, 1991).

Thomas J. Misa is assistant professor of history at Illinois Institute of Technology. He has published empirical studies on the history of chemical engineering and transistor technology as well as analytical studies of technological determinism and risk; he is a contributing editor of *Science, Technology & Human Values*. His forthcoming book on the rise of American steelmaking between 1865 and 1925 analyzes the interaction among scientific knowledge, technological processes, and industrial structures.

Michael Mulkay is professor of sociology at the University of York. He has written on art, death, and humor as well as on science. Over the years, his approach to the study of science has changed considerably, as illustrated by his most recent book, *Sociology of Science: A Sociological Pilgrimage* (Open University Press and Indiana University Press, 1991).

Trevor Pinch is associate professor in the program on Science, Technology and Society at Cornell University. His current research interests include the sociology of scientific instrumentation, the sociology of technology, and the rhetoric of selling, sales promotion, and advertising. He is the author of *Confronting Nature: The Sociology of Solar Neutrino Detection* (Reidel, 1986).

Index

Abidjan (Ivory Coast), 217–218
Abobo (Ivory Coast), 217
Actions, 236, 255n.1, 256n.11, 259,
    260–261, 304
Actor-network model/approach, 12,
    18–19, 24, 25–26, 110, 201
Actors, actants, 42, 69, 206, 208, 214,
    223n.3, 256n.11, 257n.15, 259, 262,
    299
  and closure, 110–111
  figurative and nonfigurative, 240–
    241
  historical accounts of, 76–77
  institutional, 46, 47
  nonhuman, 42, 231–236, 257–
    258nn.22, 23; 259, 261–263
  and objects, 13, 207, 220–221
  problem definition by, 140–141
  radioactive waste issue, 144, 166–167
  and resources, 302–303
  and structures, 26, 293–294, 300–302
Advertising, for fluorescent lamps,
    89–90
AEC. See United States Atomic Energy
    Commission
A.E.I.C. See Association of Edison
    Illuminating Companies
Aerospace industry, 22. See also Aircraft
Africa, photoelectric lighting in,
    209–211. See also Ivory Coast;
    Senegal
Aircraft, 23, 43(table), 44, 52n.8
  contracts for, 30, 31–34
  design of, 26, 27–28, 52nn.5, 6
  engines for, 33–35
  and technological change, 46–47
Aircraft industry, 24, 26–27, 106
Air Ministry (U.K.), 32
Air Staff (U.K.), 27–28, 30, 32, 33, 42,
    45

Amalgamation of vested interests,
    94–95, 304
American Association of Petroleum
    Geologists, 165
American Railway Review (journal), 122
Amortization of vested interests, 94,
    102n.56
Andy Falls in Love (film), 187
ANL. See Argonne National
    Laboratory
*Annual Celebration of the Schoolchildren of
    Newark N.J.* (film), 187
Anthropology, 11, 12
Anthropomorphism, 235–236, 241, 243
Antiprograms. See Programs
Antitrust Division. See United States
    Department of Justice
Argonne National Laboratory (ANL),
    144
Armat, Thomas, 185
Artifacts, 2, 75, 225–226, 247, 254,
    255–256n.5, 265
Artificial intelligence, 221
Ascription, 262
Association chains, 263(fig.), 264
Association of Edison Illuminating
    Companies (A.E.I.C.), 79
  Lamp Committee, 80, 82, 91
Association of Steel Manufacturers, 124
Atomic Energy Commission (AEC). See
    United States Atomic Energy
    Commission
Australia, 37–38
Automobiles, 56, 188, 223n.2, 249–250
Awareness contexts, 70–71

BAAS. See British Association for the
    Advancement of Science
BAC. See British Aircraft Corporation
Barriers, 297–298, 304–305

Mobility, upward, 188
Mobilization, 47, 48(fig.)
Montclair (N.J.), 187, 188
Moore tube, 82
Morality, 216, 219, 225, 227, 232, 233, 253
Morgan, J. Pierpont, 136
Motion picture industry, 185
Motion Picture Patents Company (MPPC), 186–187, 190, 192
Motion pictures, 12, 107, 197n.14, 198n.29
  class bias in, 187–189
  distribution of, 186–187
  production of, 189–190
  Thomas Edison's involvement in, 176–177, 178, 183–185, 191–193
Moza, 63
MPPC. *See* Motion Picture Patents Company
Mueller, J. E., 88, 89, 90, 94, 95, 96
Mushet, Robert, 123

NAS. *See* United States National Academy of Sciences
Nasmyth, James, 118
National Committee on Radiation Protection, 165
National Evaluation Group (CASPE), 279, 281, 288nn.17, 28
  role of, 282–285
National Health Service (NHS) (British), 203, 271, 278, 283, 299
  economics of, 268, 272
  management in, 274–275
National Reactor Testing Station (NRTS), 144
Navigation, Portuguese, 110
Negotiation, 106, 208–209. *See also* Negotiation space
Negotiation space, 21–22, 26, 42, 74n.61, 301, 305
Neilson, James, 118
Nela Park agreement, 88–89, 90, 92, 95, 96, 97
Networks, 13, 17, 21, 206, 209, 223n.1, 224n.12, 262, 303. *See also* Global networks; Local networks; Actor-network model
  electricity, 214, 215–216, 217–219, 220, 221–222, 224n.11
  heterogeneity of, 46–49
  intermediaries between, 41–42
  social, 135, 165

New Jersey Locomotive Works, 122
New York World's Fair, 82, 83, 92
NHS. *See* National Health Service
Nickelodeons, 185, 188, 301
Nimitz Highway, 1, 2–3
Noble, David, 53
Nonfigurative. *See* Figurative
North American Autonetics, 32
North Chicago Rolling Mill Company, 123
Northern States Power Company, 84, 96
NRTS. *See* National Reactor Testing Station
Nuclear deterrence, 23
Nuclear energy, 152
  Atomic Energy Commission views of, 146–150
  bureaucracy of, 142–144
  installations, 156, 159–160, 168n.5
  in United States, 141–142
Nuclear weapons, 154, 169n.14

Oak Ridge National Laboratory (ORNL), 144, 157, 159, 165, 169nn.15, 16
Obduracy. *See* Stabilization
Objects, 57
  and actants, 207, 220–221
  characteristics of, 207–208
  design and use of, 208–209
  stabilization of, 221, 224n.13
Obligatory point of passage, 31, 33, 42, 46, 48(fig.), 294, 297, 304
Oil fields, 57
Oil industry, 106
  advances in, 66–67
  exploration in, 57–66, 68–70
Olympus engine, 31–32, 34
Operational Requirement (OR) 343, 28, 30
Operational Requirements Branch (RAF), 22, 23, 25
Orford Nickel and Copper Company, 128
ORNL. *See* Oak Ridge National Laboratory
OR 343. *See* Operational Requirement 343

PACTs. *See* Planning Agreements with Clinical Teams
Paradigm, 250, 251, 263(fig.)
Paris, 227, 246